Liquid Crystals and Their Applications

Collection Editors

Ana Almeida
Pedro Marques de Almeida
Ana Catarina Baptista

MDPI

Basel • Beijing • Wuhan • Barcelona • Belgrade • Novi Sad • Cluj • Manchester

Collection Editors
Ana Almeida
Department of Chemistry
NOVA School of Science and Technology
Caparica
Portugal

Pedro Marques de Almeida
Mechanical Engineering Department
Polytechnic University of Lisbon
Lisbon
Portugal

Ana Catarina Baptista
Materials Science Department
NOVA School of Science and Technology
Caparica
Portugal

Editorial Office
MDPI AG
Grosspeteranlage 5
4052 Basel, Switzerland

This is a reprint of the Topical Collection, published open access by the journal *Crystals* (ISSN 2073-4352), freely accessible at: https://www.mdpi.com/journal/crystals/topical_collections/P93RC22I5G.

For citation purposes, cite each article independently as indicated on the article page online and as indicated below:

Lastname, A.A.; Lastname, B.B. Article Title. *Journal Name* **Year**, *Volume Number*, Page Range.

ISBN 978-3-7258-5613-8 (Hbk)
ISBN 978-3-7258-5614-5 (PDF)
https://doi.org/10.3390/books978-3-7258-5614-5

Cover image courtesy of Ana Almeida

Contents

About the Editors

Ana Almeida

Ana Almeida is a Researcher at the Portuguese Centre for Sustainable Chemistry LAQV-Requimte, awarded through a competitive international call by the Portuguese Foundation for Science and Technology (FCT). She is also a collaborator at CENIMAT | i3N (Centre of Materials Research/Institute for Nanostructures, Nanomodelling and Nanofabrication) at NOVA School of Science and Technology, Caparica, Portugal. Her research focuses on biomimetic materials, particularly membranes, fibre-based systems, and stimuli-responsive materials inspired by nature. She has developed extensive expertise in liquid crystals, particularly cellulose-based crystalline and photonic systems, investigating their self-assembly, optical properties, and practical applications. Her work integrates sustainable practices with bioinspired design to create advanced materials for technological and medical uses. She is a founding and board member of the Portuguese Liquid Crystal Society and member website editor of the International Liquid Crystal Society (ILCS). Additionally, she is a member of the International Society for Optics and Photonics (SPIE) and AMONET, the Portuguese Association of Women in Science. She also supervises PhD and master's students and teaches courses in biomimetic and polymeric materials. She actively reviews for international journals and participates in academic societies and conferences. Her scientific contributions have been recognized with several awards and fellowships.

Pedro Marques de Almeida

Pedro Marques de Almeida is an Associate Professor with Habilitation in the Mechanical Engineering Department of ISEL/IPL. He is a member of the Soft and Biofunctional Materials Group of the CENIMAT | i3N (Centre of Materials Research/Institute for Nanostructures, Nanomodelling and Nanofabrication), where he is the head of the CENIMAT NMR laboratory. He is also a member of the Unit for Innovation and Research in Engineering (UnIRE), located at ISEL, where he is President of the Scientific Council. He received a PhD and a Habilitation in Materials Science and Engineering, with main research interests focused on polymeric and mesomorphic materials, complex fluids, solid-state NMR, including MAS, diffusion, MRI and Rheo-NMR, liquid crystals, and liquid crystal applications, cellulose, natural materials, additive manufacturing and biomaterials. He has already supervised several Master's students and concluded the supervision of three PhD students, all in Materials Science and Engineering. He currently supervises one PhD student who is studying to obtain her PhD in Physics. He co-organized several international conferences mainly in the area of soft matter, liquid crystals and photonics. He has published more than 90 scientific papers, having a Scopus h-index of 25. He has participated in several national and European research projects, two of which he was the leader. Also, he leads Work Group 2 of the COST Action CA21159 (PhoBioS), exploring light-biointerface interactions for new electronic materials and devices. He is a Member of the American Physical Society and the International Liquid Crystal Society, as well as a founding member and Scientific Secretariat of the Portuguese Liquid Crystal Society.

Ana Catarina Baptista

Ana Baptista is an Assistant Professor at the NOVA School of Science and Technology (NOVA FCT) and a Full Member of CENIMAT | i3N (Centre of Materials Research/Institute for Nanostructures, Nanomodelling and Nanofabrication). Her research focuses on the design and development of polymer-based smart and functional materials for emerging applications in health, bioelectronics, and wearable technologies. Her work explores the interface between materials science, electronics, and biotechnology, emphasizing sustainable and biocompatible approaches to energy storage, sensing, and controlled drug delivery. Ana has led research on cellulose-based supercapacitors, biocompatible bio-batteries, electrospun glucose-sensitive membranes, electronic textiles, and electrically responsive drug-release systems. As Principal Investigator of the All-FIBRE project (2021–2024), she coordinated the development of fibre-shaped photovoltaic supercapacitors and solar cells capable of harvesting and storing energy within the same lightweight textile structure. Over the past decade, Ana has authored more than 30 scientific publications, supervised over 40 master's students, and currently mentors two PhD candidates. She has been involved in eleven funded research projects, serving as Principal Investigator or Co-Principal Investigator in four of them, and has also contributed to several national and international collaborations in energy generation and storage, functionalization of paper and polymeric substrates, and 3D printing of biocompatible polymer and ceramic materials for medical implants. Beyond research, she plays an active role in academic management at NOVA FCT and serves as an Associate Editor for the journal *3D Printing in Medicine*.

Preface

Liquid crystals (LCs) are among the most impressively varied classes of soft condensed matter, interpolating order and fluidity to produce a fascinating array of structural, optical, and dynamic phenomena. Their unique mesophases, neither solid nor liquid, give them tuneable anisotropy, external stimulus sensitivity, and self-organization on more than one length scale. As liquid crystals first became widely known through their use in display technology, their significance has evolved significantly beyond information imaging and has encompassed modern photonics, materials science, chemistry, biology, and energy systems. The Topical Collection "Liquid Crystals and Their Applications" published in *Crystals* was designed to reflect this diversity and to highlight modern frontiers in LC research. Its scope varies from elementary research on molecular ordering, electro-optical interaction, and phase transitions to technological developments in energy conversion, microfluidics, and biomimetic materials. The present Reprint gathers scientific research that demonstrates how recent studies continue to extend the boundaries of LC potential through advances in material design, experimental methodology, and modelling approaches.

The general aim of this volume is to merge contributions demonstrating how liquid-crystalline systems can be implemented as active functional platforms in chemistry, physics, engineering, and biology. Each contribution addresses a various facet of LC science, from novel molecular architectures to dynamic optical effects and hybrid materials for photonic and electronic devices, due to the multidisciplinary character of the field. Combining these studies, there is a coherent tale of engineering, examining, and directing molecular order towards technology-targeted and environmentally friendly solutions.

In "Light-Induced Heating of Microsized Nematic Volumes", Dmitrii Shcherbinin and colleagues offer a stunning experimental study on thermomechanical stresses in hybrid-aligned nematic microcells exposed to laser light. Using advanced computer-vision-based image analysis, the authors reveal isotropic domain formation and concentric ring textures due to local temperature gradients. Their observations define two regimes, "fast" and "slow" heating modes, corresponding to direct laser absorption and heat diffusion through indium-tin-oxide layers, respectively. This work not only presents quantitative metrics for texture evolution but also underlying physical insights into director deformations and thermal fields, the foundations of future breakthroughs in optically driven microfluidics and soft actuators.

Paweł Szubert and Stanisław A. Różański discuss the potential of nematic LCs to enhance renewable-energy devices in "The Effect of Nematic Liquid Crystal on the Performance of Dye-Sensitized Solar Cells". Their research indicates that doping standard iodide electrolytes with the nematic compound 4-n-pentyl-4′-cyanobiphenyl (5CB) controls current-voltage behaviour and photovoltaic performance. Orientational ordering of LC molecules near TiO_2 photoanodes affects charge transport and electrolyte stability, with a beneficial concentration optimized to maximize performance before insulating mechanisms dominate. This work connects soft-matter self-assembly to energy conversion, which opens new possibilities for LC-based hybrid electrolytes in dye-sensitized and perovskite solar cells.

In "Maintaining Glycerol-Based Hexagonal Structures by Crosslinkers for High Permeability Nanofiltration", Senlin Gu, Luke A. O'Dell, and Lingxue Kong describe how chemical crosslinking maintains glycerol-based hexagonal LC phases to form nanofiltration membranes. The authors demonstrate that the introduction of selective crosslinkers immobilizes the ordered LC nanostructure against water-induced swelling to generate highly permeable but resilient membranes. Their findings link molecular-level ordering in lyotropic LCs to productive separation technologies and propose new design routes for sustainable, high-performance filtration materials.

Hyeon Seong Hwang et al. present "Broadband Near-Infrared Reflective Film from Stacked Opposite-Handed Chiral Liquid Crystals with Pitch Gradients". The authors design and produce multilayer cholesteric LC films that combine opposite-handed helices with spatially varying pitches, inducing broadband reflection in the near-infrared regime. Their work integrates structural colour control and thermal regulation, proposing avenues in smart coatings and efficient optical devices. The paper nicely illustrates how chirality engineering and pitch gradient modulation can unleash the photonic potential of LC materials.

In "Light-Induced Higher-Order Bragg Resonance in Heliconical Cholesteric Liquid Crystals", Ashot H. Gevorgyan and Francesco Simoni discuss the nonlinear optical response of heliconical cholesteric structures to light. They notice the emergence of higher-order Bragg resonances, which occur owing to the helical pitch modulation and can be controlled by external fields. This discovery extends the photonic tunability of cholesteric phases, providing an explanation for how complex helical geometries are able to generate dynamic and tuneable colour responses for photonic crystals and optical switches.

Izabela Śliwa, Pavel V. Maslennikov, and Alex V. Zakharov provide "Dissipation of Energy in a Compressible Nematic Microvolume under Effect of a Temperature Gradient". From earlier theoretical studies, they examine mechanisms of dissipation of internal energy in nematic volumes subjected to thermal gradients. According to their theory, director deformation coupling with compressibility generates viscous and thermomechanical losses. This paper fosters the fundamental thermodynamics of liquid crystals to microscale heat transfer and energy conversion in anisotropic fluids.

Aleksey Kudreyko, Vladimir Chigrinov, and Arina Perestoronina study photoalignment phenomena at the interface of films of azo dyes and planar-aligned nematic liquid crystals in "Photoinduced Interactions in Thin Films of Azo Dyes and Planar-Aligned Nematic Liquid Crystal". Light-induced reorientation of molecules producing stable and controllable alignment patterns is demonstrated through their experiments. The findings elucidate primary processes of light-induced molecular ordering and hold promise for reconfigurable liquid crystal devices, holographic elements, and advanced optical storage devices.

Anastasiia Svanidze and co-workers submit "The Effects of Network Architecture on the Photomechanical Performance of Azo-Acrylate Liquid Crystal Elastomers". The work explores how changes in crosslink density and mesogenic content impact photomechanical deformation in LC elastomers. The authors demonstrate that tailored network architectures are the reason for enhanced photomechanical efficiency and mechanical reversibility and provide valuable design guidelines for adaptive optical devices and responsive soft actuators. This work bridges the areas of polymer chemistry and LC physics to create new classes of photoresponsive materials.

In "Review of Angular-Selective Windows with Guest–Host Liquid Crystals for Static Window Applications", Seung-Won Oh and Chan-Heon An summarize the basics and recent progress of angular-selective glazing systems based on guest–host LCs. They describe how doped LCs' orientation-dependent scattering and absorption are used to manage daylight and solar heat gain passively without active control. The extensive review bridges optical anisotropy and architectural energy efficiency, outlining paths toward passive smart-window solutions.

Finally, Qi Wu, Hongxia Zhang, Dagong Jia, and Tiegen Liu present "3D Optical Wedge and Movable Optical Axis LC Lens", a paper introducing a new LC-based optical wedge that can modulate the focus in three dimensions. With precise control over LC alignment and electrode pattern, the authors achieve continuous modulation of the optical axis and focal position. This innovation paves the way for compact, tuneable optical devices in augmented reality, adaptive imaging, and photonic beam steering.

Collectively, these advances illustrate the adaptability of liquid crystals across fields. From the optical control and micro-scale thermal transport to energy harvesting and biosensing, LCs have an unparalleled capacity to combine order, dynamics, and functionality. Out of this Reprint comes the picture of a field in perpetual transformation, one fuelled by competition between fundamental comprehension and technological advancement. The merging of advanced modelling, precision optics experiments, and new materials synthesis indicates a trend toward multifunctional LC systems responsive to light, heat, electric fields, and chemical stimuli.

There are some common strands. Firstly, the union of structure and function, the ability to engineer molecular and mesoscale order to control macroscopic properties, is the thread that runs throughout all the papers. Second, multiphysics coupling, where electrical, optical, and thermal phenomena exist and are coupled together, is what enables most of the discoveries outlined here to occur. Third, and most fundamentally, there is a shared ethos of sustainability and cross-disciplinary relevance: LC research today not only pushes optics and electronics but also renewable energy, biocompatible systems, and smart materials with small environmental footprints.

The successes documented in this Topical Collection once again attest to the fact that liquid crystal science is far from mature. New classes of molecules, such as ferroelectric nematics, ionic liquid crystals, and lyotropic bio-mesogens, are extending the intellectual boundaries of the discipline. New experimental methodologies, such as laser-induced microdynamics and computer-vision-based analyses, are illuminating with fresh insights into complex LC behaviour under stimulation. The convergence of these approaches heralds an exciting era in which liquid crystals will continue to be the fertile ground for innovation and imagination.

The Guest Editors offer their sincere thanks to all the authors whose work is included in this Reprint. Their creativity, intelligence, and tenacity have collectively advanced knowledge of liquid crystalline systems and demonstrated their broad applicability in science and technology. Each of the papers in this volume reflects both the heterogeneity and the cohesion of the LC research community: heterogeneity of method and application, cohesion in attempting to comprehend and exploit ordered soft matter for the benefit of society. We also wish to thank the reviewers and editorial staff of *Crystals* for upholding the scientific integrity and visibility of this work.

Ana Almeida, Pedro Marques de Almeida, and Ana Catarina Baptista
Collection Editors

crystals

MDPI

Article

3D Optical Wedge and Movable Optical Axis LC Lens

Qi Wu [1,2], Hongxia Zhang [1,2,*], Dagong Jia [1,2] and Tiegen Liu [1,2]

[1] School of Precision Instrument and Opto-Electronics Engineering, Ministry of Education, Tianjin University, Tianjin 300072, China; qiwu@tju.edu.cn (Q.W.); dagongjia@tju.edu.cn (D.J.); tgliu@tju.edu.cn (T.L.)
[2] Key Laboratory of Optoelectronic Information Technical Science, Ministry of Education, Tianjin University, Tianjin 300072, China
* Correspondence: hxzhang@tju.edu.cn

Abstract: Current liquid crystal (LC) lenses cannot achieve lossless arbitrary movement of the optical axis without mechanical movement. This article designs a novel bottom electrode through simulation and optimization, which forms a special LC lens with an Archimedean spiral electrode, realizing a 3D LC wedge and an arbitrarily movable LC lens. When only the bottom electrode is controlled, it achieves a maximum beam steering angle of 0.164°, which is nearly an order of magnitude larger than the current design. When the top and bottom electrodes are controlled jointly, a 0.164° movement of the lens optical axis is achieved. With focal length varies, the movement of the optical axis ranges from zero to infinity, and the lens surface remains unchanged during movement. The focus can move in a 3D conical area. When the thickness of the LC layer is 30 μm, the fastest response time reaches only 0.635 s, much faster than now.

Keywords: LC lens; movable optical axis; 3D LC wedge; electrode design; unchanged

1. Introduction

Nowadays, tunable optical wedges and optical axis motion (OAM) lenses are capable of achieving optical image stabilization (OIS) [1–4], auto-focus [5], super-resolution imaging [6–8], 3D reconstruction [9–11], and optical tweezers [12,13], etc. Meanwhile, the movement of traditional solid lenses needs mechanical movement, thereby increasing power consumption, occupying space, and presenting low tuning efficiency [14].

Traditional adaptive optics—a technology aimed at improving the performance of optical systems—can indeed achieve beam steering [15] and focusing [16–18], but it requires the control of multiple mirrors to construct a deformable mirror surface, resulting in high costs and complex structure. A comprehensive system is also indispensable for real-time feedback, including a wavefront detector, wavefront controller, and wavefront corrector [19], which leads to a large size. Therefore, it is usually used in fields such as astronomy, which requires high precision and multi-discipline. However, LC lenses have significant advantages in terms of miniaturization, are lightweight, and have a low cost when applied voltage of less than 10 V [20]. The millimeter-level dimensions and micron-level thickness render the LC lens portable and suitable for mobile phones and other small devices, making it perfectly convenient in life.

LC devices are the typical representative of novel adaptive optics and function as a phase modulation element. The researches encompass LC beam deflectors [21,22] and LC zoom lenses [23,24]. The beam deflectors are classified into refractive wedges and diffraction gratings [22]. The former generates a small refractive angle yet incurs a relatively small loss of light energy, while the latter exhibits a large angle but has multiple diffraction orders, resulting in a more significant light loss. The research on refractive LC wedges is gradually becoming more comprehensive. Love [25] and Hands [26] have achieved beam steering in 1D with a 0.01° steering angle and in 2D with a 0.024° steering angle. Kotova [27] employed a modal four-electrode device to observe the 3D continuous optical deflection experimentally. Nevertheless, the deflection theory was not improved until Xu [28], accomplishing 0.048° beam deflection.

The LC zoom lens, initially proposed by Sato and Ye [20,23,29–31], is capable of adjusting the focal length and optical axis without mechanical movement. Nevertheless, the OAM lens exhibits poor performance and significant aberrations, and it has difficulty maintaining shape during movement. Meanwhile, the modal electrodes produce an uneven and unstable voltage distribution. With the advancement of LC pattern electrodes with micro-structured [32–36], the voltage across the aperture can be precisely characterized by the expression. However, independent research on LC devices has resulted in low utilization efficiency of layer electrodes.

Therefore, through multi-simulations and optimizations, this paper proposes a novel LC lens with a movable optical axis that can achieve the movement of focus in 3D space without mechanical movement. It possesses the merits of adjustable focal length, lightweight, compact structure, no mechanical movement, and low power consumption. This paper elaborates on the design principle, working principle, and driving method of the electrodes and fabricates a liquid crystal lens with a liquid crystal layer of 30 μm. The interference fringes and positions during the movement process were recorded, and they were compared with the theoretical values.

2. Principle

2.1. Electrode Structure and Process

For the OAM-LC lens, the device structure is presented in Figure 1a, comprising five parts: glass, top electrode, LC layer, bottom electrode, and glass. The top electrode, as shown in Figure 1b, is an Archimedean spiral with equal spacing, which is designed based on the patterned electrode theory to achieve low aberration focusing [37]. The bottom electrode is designed based on vector synthesis and modal electrode theory [38], generating a planar phase difference to achieve beam steering and focus movement. The bottom electrode structure is shown in Figure 1c. When the top and bottom electrodes are jointly controlled, as depicted in Figure 1d, the focus can move within the 3D space, and its movement range lies within the conical area presented in Figure 1e. The dioptre power is within ±6D, the vibration of the l optical axis ranges from 0.164°, and the movement of the optical axis can reach infinity.

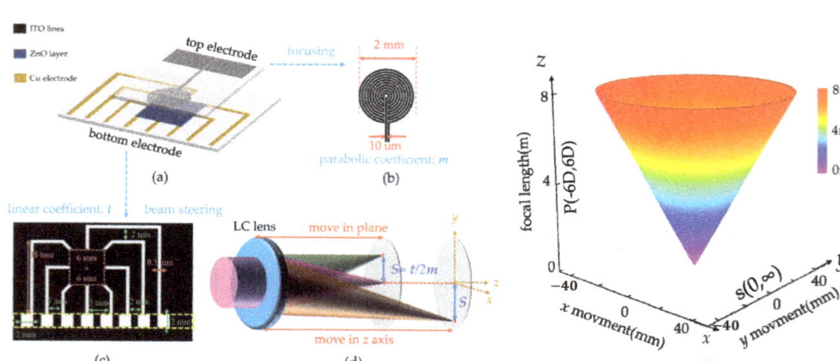

Figure 1. (**a**) 3D structure diagram of the device, (**b**) top electrode with electrode wire and gap are both 10 um; (**c**) bottom electrode, (**d**) movement of OAM LC lens focus, (**e**) focus moving area in 3D space (The color represents the focal length).

It is worth noting that the bottom electrode is composed of copper (Copper) and zinc oxide (Zinc Oxide); hence, the size and brief manufacturing process of the electrode must be explained. The size of the glass is 25 mm × 20 mm. Since there is an extremely negligible voltage drop across the copper wires, it is sufficient to ensure that they can be connected.

The current design of copper wires features a width of 0.5 mm, with the central Zinc Oxide region being 6 mm×6 mm. The total radius of the top electrode is 1 mm, as depicted in Figure 1b, with electrode widths and gaps both 10 µm.

We simulated the voltage distribution by COMSOL, and a perfect oblique planar phase difference was generated in the central 2 mm × 2 mm region within the ZnO area, as shown in Figure 2a,b. This implies that the top electrode generates a parabolic voltage distribution while the bottom electrode generates a planar voltage distribution, as shown in Figure 2c. The voltage difference within the LC cell is determined by the top and bottom electrodes.

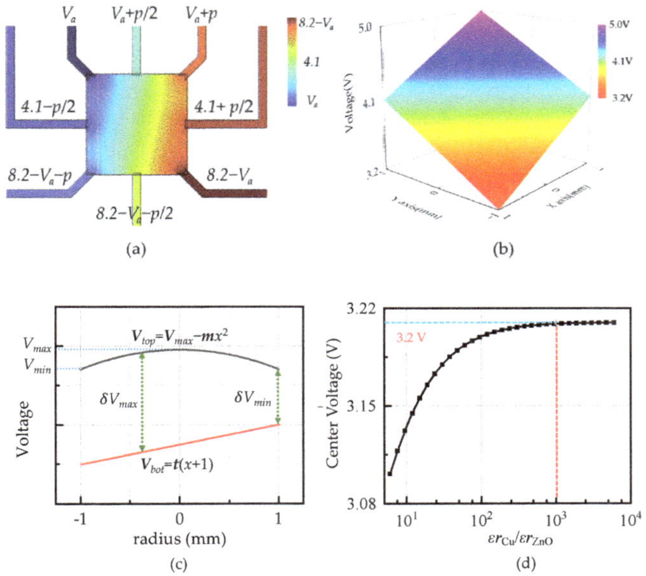

Figure 2. (**a**) The whole region with V_a is the minimum input voltage of the eight copper electrodes; p is the voltage increment in the X direction; (**b**) the core region of the bottom electrode; the color bars in (**a**,**b**) are used to display the voltage within the area (**c**) the voltage difference between the top and bottom electrode. the black line represents the top layer voltage distribution, the red line represents the bottom layer voltage distribution, and the green dashed line represents the voltage difference between the upper and lower electrodes; (**d**) the variation in center voltage with the sheet resistance (εr) ratio, the red and blue lines respectively represent the sheet resistance ratio and voltage value when the voltage is exactly stable.

The bottom electrode is developed using two photolithographic steps for the deposition of high-resistance copper and zinc oxide films. The rotating speed of the spin-coated photoresist is 6000 RPM. The deposition duration for copper and zinc oxide high-resistance films is approximately 30 min and 35 min, respectively. The thickness of the zinc oxide film is approximately 25 nm. The resistance εr_{cu} of the copper electrode is approximately 300 Ω, and the ε_{rZnO} of the zinc oxide region is approximately 1 MΩ. The resistance ratio between the two is approximately 10^3. In accordance with the simulation in Figure 2d, when the sheet resistance ratio (ε_r) between the two exceeds 10^3, the copper will scarcely generate a voltage drop.

2.2. Function and Working Principle

This structure can realize an optical-axis-movable LC lens (OAM LC lens), which is a highly significant function. OAM lenses can improve the imaging performance of the optical system through adaptive zooming and simultaneously eliminate image shifts induced by the optical axis shift and tilt of the camera. Hence, the system holds potential applications in optical image stabilization (OIS). Thus, the system can realize multiple functions, and the specific functions and theoretical derivations will be expounded in detail in the subsequent sections.

3D Optical wedge: According to the simulation in Figure 2b, the center of the bottom electrode can generate a planar phase profile, which means an optical wedge can be made. Moreover, the steering angle θ_w and azimuth angle θ_a are controlled by two parameters, V_a and p. The θ_w and θ_a are expressed as:

$$\theta_a = \arctan(\frac{N\lambda}{Diameter}) = \arctan(\frac{\Delta n \cdot h_{LC}}{Diameter}) \quad (1)$$

$$\theta_a = \arctan(\frac{\partial \Delta n}{\partial x} / \frac{\partial \Delta n}{\partial y}) \quad (2)$$

N: the number of stripes, λ: laser wavelength; Δn: refractive index difference; h_{LC}: LC layer thickness.

Zoom imaging: The voltage distribution of the top electrode and bottom electrode are parabolic and planar, respectively, as we can see in Figure 2c.

The voltage difference within the aperture can be expressed as Equation (2):

$$v = V_{max} - m \cdot x^2 - t(x+1) \quad (3)$$

m: the parabolic coefficient (V·mm^{-2}); t: the linear coefficient (V·mm^{-1}); V_{max}: the maximum voltage on the top electrode; V_{min}: the minimum voltage on the top electrode; x (mm) is the position of a point within the aperture. To generate parabolic and planar phase profiles, all voltage difference values within the aperture must fall within the linear response region of the LC material [39]. If $t = 0$, the voltage difference within the aperture is a parabolic distribution to be a lens with a small wavefront RMS error (<0.05 λ).

OAM lens: by transforming the Equation (3), the voltage difference can be expressed as Equation (4):

$$v = (V_{max} - t + \frac{t^2}{4m}) - m(x + \frac{t}{2m})^2 \quad (4)$$

From Equation (4), the axis of the parabolic function has shifted, but its shape has not changed with parabolic coefficient m remains. This implies that when the top and bottom electrodes are jointly controlled, they can be OAM LC lenses and remain unchanged during the movement of the optical axis. The movement of the optical axis is obtained.

$$s = \frac{t}{2m} \quad (5)$$

s: the movement of the optical axis of the OAM lens. From Equation (5), the optical axis of the lens can move to any position, but if it is too far away, it is meaningless for the 2 mm × 2 mm lens. Therefore, achieving movement within the 2 mm × 2 mm aperture is sufficient.

Offset image shifts caused by optical axis vibration: When photographing images of distant objects, minor shifts in the imaging system induced by external forces will give rise to the vibration of the lens. If the conventional solid lens is still employed, the imaging position will undergo displacement. The prevailing OIS technology is cropping the image. Nevertheless, by utilizing the aforementioned OAM lens, optical image stabilization can be accomplished without any information loss and mechanical movement through the lens moving in the opposite direction. Its working principle is shown in Figure 3a.

The offset of the image can also be explained by the following expression. The image movement is equal to the steering angle multiplied by the focal length. The focal length of the LC lens is expressed as:

$$f = \frac{r^2}{2d \times \Delta n} \tag{6}$$

The offset of the image is:

$$\Delta h = f \tan\theta = \frac{t}{2m} = s \tag{7}$$

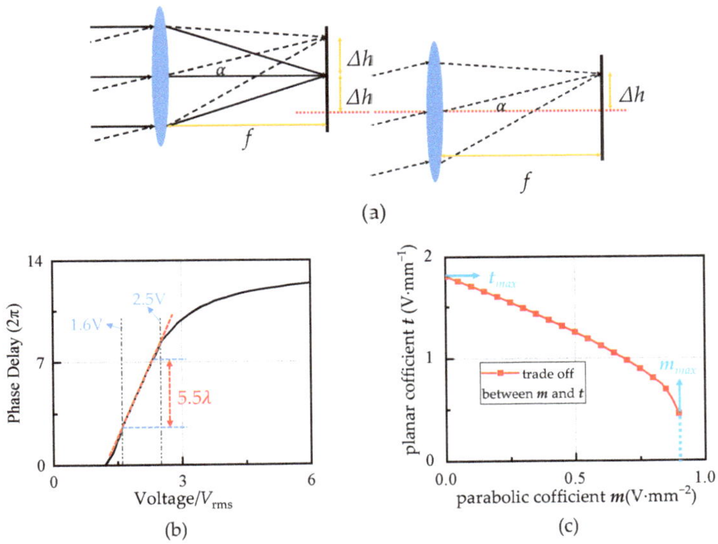

Figure 3. (**a**) Schematic of image shift caused by optical axis vibration and OAM lens eliminate the shift (the black line represents the light path, while the red line represents the movement of the lens); (**b**) extracted phase delay of the LC layer and the linear response range of the LC material, the black line represents the response of the LC, and the red dashed line represents the linear response region of the LC; (**c**) the red line is the trade-off between m and t, blue line represents the range of the m and t.

By combining Equations (5) and (7), the offset of the image is exactly equal to the shift of the lens optical axis, and they compensate for each other.

Overall, this is why the lens can be used for OIS. It can eliminate parallel movement of the optical axis and compensate for image offset caused by optical axis vibration. This shows the multifunctionality of the OAM lens.

2.3. Driving Voltage Condition

Within the aperture, the voltage difference must fall within the linear response region (1.6~2.5 V) of the LC material depicted in Figure 3b. Therefore, it is necessary to satisfy the condition that $\delta V_{max} - \delta V_{min} \leq 5$ V–3.2 V. Moreover, the driving voltages require rigorous calculations.

$$\delta V_{max} = V_{max} - m \cdot \left(-\frac{t}{2m}\right)^2 - t\left(-\frac{t}{2m} + 1\right) = V_{\max} + \frac{t^2}{4m} - t \leq 5 \tag{8}$$

$$\delta V_{min} = V_{max} - mr^2 - (r+1)t \geq 3.2 \tag{9}$$

δV_{max}: the maximum voltage difference in the 2 mm × 2 mm aperture; δV_{min}: the minimum voltage difference in the 2 mm × 2 mm aperture; r is the radius of the LC lens. Combining Equations (8) and (9), we can obtain the trade-off between m and t as shown in Equation (10):

$$4m^2 + (4tr - 7.2)m + (tr)^2 \leq 0 \tag{10}$$

Thus, the driving voltage range is expressed as follows:

$$V_{min} \geq 3.2 + 2tr \tag{11}$$

$$m \leq \frac{-(4tr - 7.2) + \sqrt{7.2 \times (7.2 - 8tr)}}{8} \tag{12}$$

In this article, the radius of the OAM LC lens is 1 mm. The trade-off between m and t is depicted in Figure 3c.

3. Experiment and Result

3.1. Experimental Setup

We designed two interference optical paths in Figures 4 and 5 for the experiment. The device in Figure 4 was used to record the fringes to verify the feasibility of the OAM lens. The laser wavelength is 532 nm, and the modulator is used to adjust the intensity of the laser. When the laser passes through the polarizer I, it is decomposed into two orthogonal components. The first one sees the LC ordinary refractive index, which is uniform. However, the second one shows the LC's extraordinary refractive index, which is affected by the applied voltages. The two components have phase differences; as they pass through the polarizer II, the interference fringes are captured by the CMOS camera.

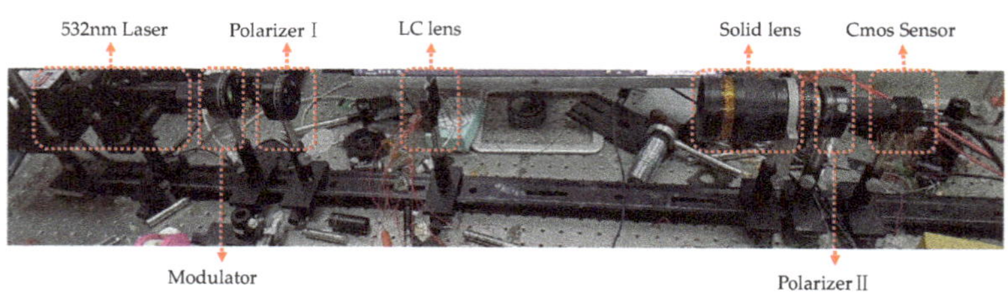

Figure 4. Orthogonal interference setup to capture the fringes.

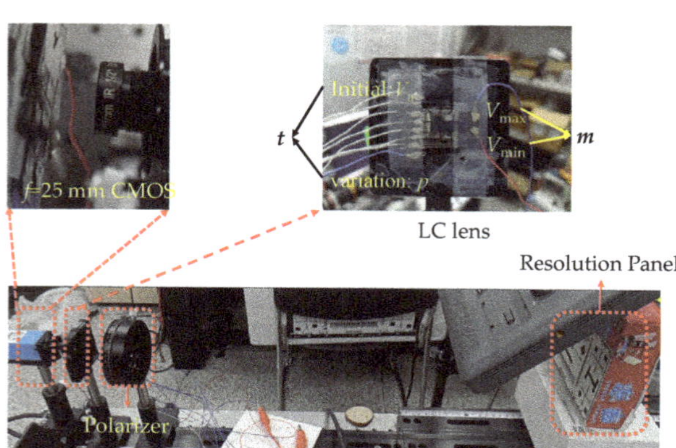

Figure 5. Image movement observation experimental setup, the prepared LC lens, and wire leads. The resolution panel is attached on the box for fixation.

The setup in Figure 5 was used to capture the image shifts caused by the OAM lens in the imaging system. We place the OAM lens in the imaging system, and the optical wedge will deflect the incident light, causing an image shift.

The prepared LC lens is shown in Figure 5. The left electrodes control the bottom voltage to achieve the OAM. The two left electrodes control the upper voltage and the focusing effect of the LC. The positive nematic LC material used in the experiment is HTW 148700-100, produced by Hecheng Display Technology Co., Ltd. from Yangzhong, Jiangsu China, whose optical birefringence $\Delta n = 0.259$. The linear response range of this LC material is measured using the method reported in [40], which is 1.6~2.5 V_{rms}. Then, the driving voltage difference falls within this range to generate a specific phase profile. The driving voltages are square waves. Therefore, 1.6–2.5 V_{rms} is equivalent to 3.2–5.0 V_{pp}, and the voltage mentioned below in the article is the V_{pp}. The frequency of the driving voltage is all 1 kHz in this work.

3.2. Analysis of Light Deflection Experiment

The driving voltage is related to two parameters, V_a and p. V_a determines the azimuth angle θ_a, and p determines the steering angle θ_w. The interference fringes of the 3D optical wedge are recorded in Figure 6a–f with different driving voltages listed in Table 1.

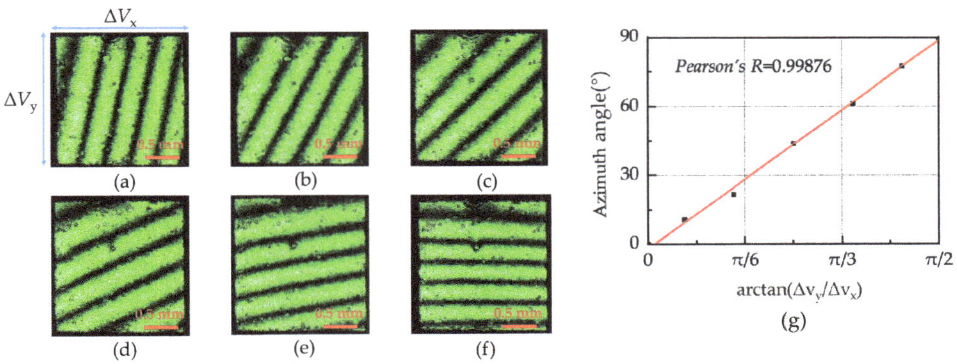

Figure 6. **(a–f)** The 3D spatial light deflection effect of the LC wedge; **(g)** the relationship between the azimuth angle of the LC wedge and the voltage distribution on both sides.

Table 1. The driving voltage and parameter of the different azimuth angle.

	(a)	(b)	(c)	(d)	(e)	(f)
V_a	0.8 V	0.8 V	0.8 V	0.8 V	0.8 V	0.8 V
p	1.1 V	2.2	3.3	4.4	5.5	6.6
θ_a	77.6	61.3	43.8	26.3	10.6	0
θ_w	0.17°	0.17°	0.17°	0.17°	0.17°	0.17°

The results shown in Figure 6 demonstrated that the LC optical wedge adjusts from 0–90° as the fringes with different tilt angles. There are approximately 5.5 λ variations within the aperture, which means the wedge angle is 0.0838° from Equation (1).

According to the voltage from Table 1, the ratio of voltage drops across the X (ΔV_x) and Y (ΔV_y) axis is related to the slope of the fringes. Due to the slope of the tilt fringes and the fact that the ratio may be infinite, they were processed by means of the inverse tangent function. A relationship between the azimuth angle and the ratio is plotted in Figure 6g, demonstrating a linear relationship in the form of a proportional function.

The proportional relationship means that the structure has the advantage of controlling the azimuth angle of the beam steering by applying voltage ΔV_y and ΔV_x across the electrode. The linear fit is as high as 0.999, nearly to 1.

The LC optical wedge is placed into the imaging system in Figure 5, and the practicability of the LC optical wedge is verified by the imaging resolution plate in the experiment. Figure 7e is the original state; the surrounding images are shifted by the wedge. The movement in Figure 7a–i is depicted in Figure 7j.

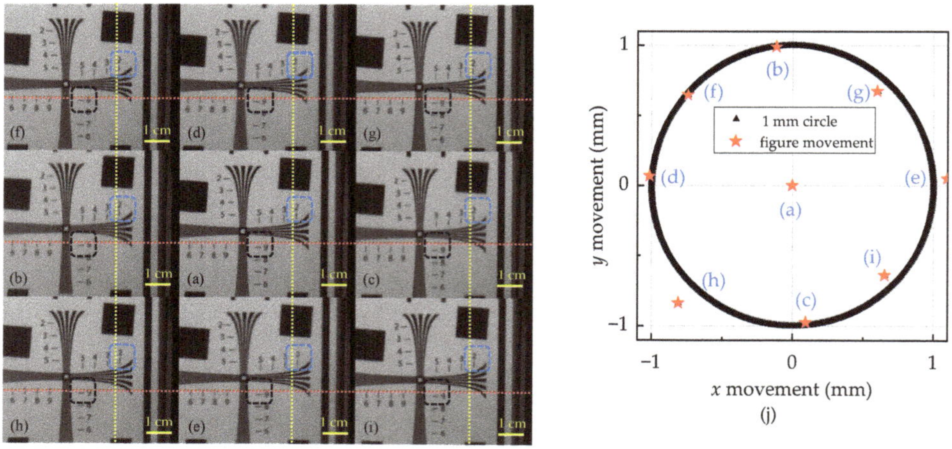

Figure 7. Camera imaging position shift under LC wedge: center image (**a**), up shift (**b**), lower shift (**c**), left shift (**d**), right shift (**e**), up left (**f**), up right (**g**), lower left (**h**), lower right (**i**). Red and yellow lines are observation aids; the blue box '2' indicates horizontal movement comparison, and the black box '9' indicates vertical movement comparison (**j**) movement of figures.

In the optical imaging system shown in Figure 5, the object distance is 700 mm, with a 25 mm lens employed, giving rise to a maximum image shift of 1 mm. According to geometric calculations, the maximum angular deviation in one direction can reach 0.164°.

3.3. Analysis of Off-Axis Focusing Experiment

It should be noted that when applying the voltage to the OAM lens, the trade-off between the m and t needs to be considered according to Equation (12), as shown in Figure 3c.

According to Equations (4) and (5), m determines the dioptre power, and m and t jointly determine the movement of the lens s, which means that the trade-off between the m and t leads a trade-off between the movement s and the dioptre power as depicted in Figure 8. The maximum dioptre power of the lens is ±6D. Once the OAM lens moves to infinity, the dioptre power is nearly 0, the movement can be calculated by Equation (6), and the fringes are parallel.

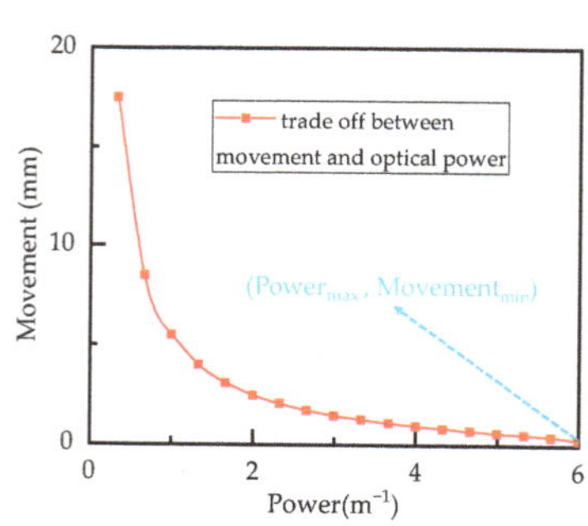

Figure 8. The trade-off between movement and dioptre power of the OAM lens.

Four experiments were proposed with different values of m and t to demonstrate the OAM lens. Here, we define S_{figure} as the real movement of the OAM lens in figures.

3.3.1. m and t at the Critical Condition

When the voltage differences within the aperture entirely fall within the linear response range of the LC material (1.6–2.5 V_{rms}), the LC lens can achieve the specific function. Hence, the critical conditions of m and n addressed in this article occur when δV_{max} = 2.5 V_{rms} and δV_{min} = 1.6 V_{rms}, as Equation (12) met the condition of equality. The values of m and t within the critical conditions are a must to guarantee that the voltage difference lies within the linear response range of the LC material. The critical condition of m and t is listed in Table 2, and the linear response region of the LC material is fully used (δV_{max} = 5.0 V, δV_{min} = 3.2 V), ensuring that at least five fringes can be observed in Figure 9.

Table 2. m and t at the critical condition.

m (V·mm^{-2})	t (V·mm^{-1})	s (mm)	S_{figure} (mm)	V_a (V)
1.70	0.10	0.029	0.03	3.73
1.59	0.20	0.063	0.07	3.37
1.48	0.30	0.101	0.11	3.00
1.37	0.40	0.146	0.16	2.63
1.25	0.50	0.200	0.22	2.27
1.12	0.60	0.268	0.27	1.90
0.97	0.70	0.361	0.38	1.53
0.45	0.80	0.500	0.52	1.17
0.45	0.90	1.000	0.98	0.80

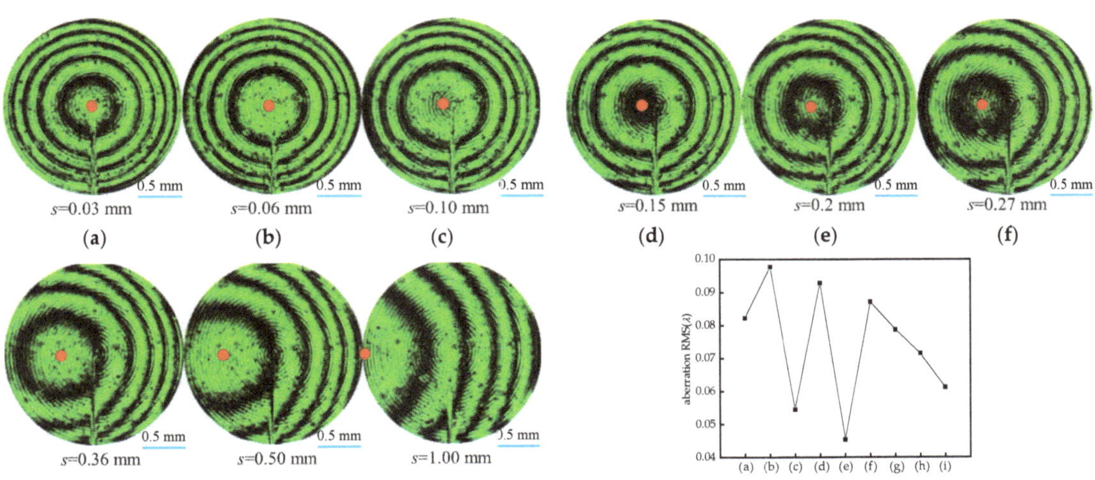

Figure 9. Interference fringes of OAM LC lens when m and t are within critical condition (**a**) t = 0.1; (**b**) t = 0.2; (**c**) t = 0.3; (**d**) t = 0.4; (**e**) t = 0.5; (**f**) t = 0.6; (**g**) t = 0.7; (**h**) t = 0.8; (**i**) t = 0.9, (**j**) the aberration RMS error, and the red dot is the center of the lens.

From Figure 9a–i, the lens shifts from the aperture center. The movement of the OAM lens S_{figure} is recorded by pixels, compared with the theoretical value s. The aberrations of the OAM lens were recorded to verify its feasibility as a lens in Figure 9j, and the aberrations were all less than 0.1 λ, which demonstrates the good imaging performance of the proposed OAM LC lens. Therefore, when the OAM lens moves unidirectionally at critical conditions, the RMS error is no longer recorded.

3.3.2. Fixed m, Gradual t, Fixed Focal Length OAM Lens with Equal Movement

Fixed m at 0.8, the maximum value of t was 0.8 from the Table 1. According to Equations (4) and (5), fixed m means fixed focal length and a gradual change in t means a gradual change in the movement of the OAM lens. s. t is gradually changed in increments of 0.2. These m and t are listed in Table 3. The interference fringes of the lens are measured by the setup in Figure 4. Accordingly, the interference fringes of the lens only exhibit movement in Figure 10a–e.

Table 3. The driving voltage and parameter of the OAM.

	(a)	(b)	(c)	(d)	(e)
V_{min} (V)	7.3	7.5	7.7	7.9	8.1
V_{max} (V)	8.1	8.3	8.5	8.7	8.9
m (V·mm^{-2})	0.8	0.8	0.8	0.8	0.8
t (V·mm^{-1})	0	0.2	0.4	0.17°	0.17°
s (mm)	0	0.125	0.250	0.375	0.500
S_{figure} (mm)	0	0.126	0.251	3.780	0.495

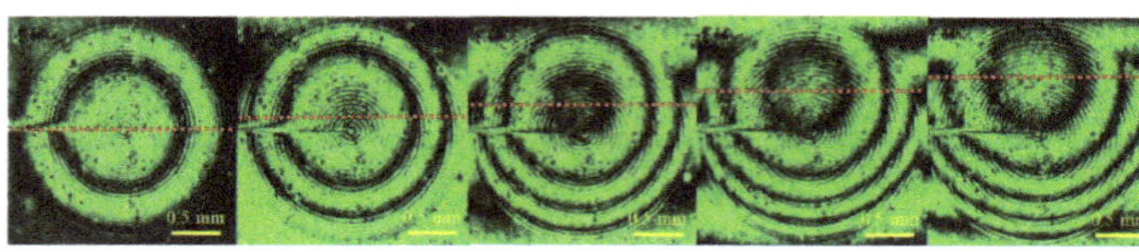

(a) (b) (c) (d) (e)

Figure 10. The interOAM lens in one direction with different movements: (**a**) initial state; (**b**) 0.125 mm movement; (**c**) 0.25 mm movement; (**d**) 0.375 mm movement; (**e**) 0.5 mm movement. The red line shows the vertical movement of the lens.

3.3.3. Changing m and t, Fixed OAM Lens with Changed Focal Length

According to Equations (4) and (5), if the optical axis of the OAM lens is to be fixed, both m and t should be adjusted simultaneously to ensure that the position of the optical axis s remains constant. In this part, three fixed optical axis OAMs were recorded in Figure 11. The movements of the OAM lens are s = 0.25 mm, s = 0.5 mm, and s = 1.00 mm. The concrete values of m and t are listed in Table 4

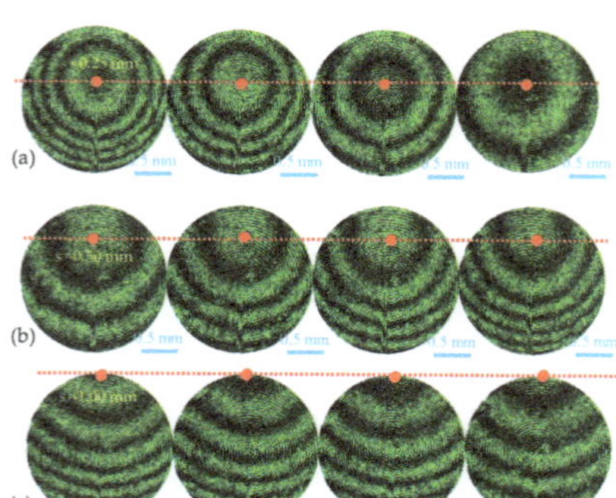

Figure 11. Interference fringes of the OAM lens with a fixed optical axis that represents the following s fixed movements: (**a**) 0.25 mm; (**b**) 0.5 mm; (**c**) 1.00 mm of the OAM lens. The red dot is the center of the lens and the red line shows the vertical movement of the lens.

Table 4. The driving voltage and parameter of the fixed optical axis.

s (mm)	m (V·mm^{-2})	t (V·mm^{-1})	V_a (V)
0.25	0.40	0.20	3.37
	0.60	0.30	3.00
	0.80	0.40	2.63
	1.00	0.50	2.27
0.50	0.40	0.40	2.63
	0.50	0.50	2.27
	0.60	0.60	1.90
	0.70	0.70	1.53
1.00	0.30	0.60	1.90
	0.35	0.70	1.53
	0.40	0.80	1.17
	0.45	0.90	0.80

When the optical axis of the OAM lens is fixed, we can observe that the m and t increase simultaneously, which means the number of fringes within the aperture increases. The red dashed line refers to the position of the center of the lens. A series of fringes demonstrated that once the optical axis of the OAM lens is fixed away from the center, the changes in the m and t can only influence the focal length of the lens from Equation (5); it can achieve zoom at a position away from the center of the lens.

3.3.4. Fixed m and t, the OMA Lens Moves within the Aperture in Multiple Directions

Finally, we demonstrated the multi-directional movement of the OAM lens within the aperture. In order to record fringes clearly, the m and t are fixed at 1.0 and 0.6, respectively. The constant movement of the optical axis is s = 0.3 mm. The interference fringes are captured during the optical axis shift in Figure 12a–e.

(a) (b) (c) (d) (e) (f)

Figure 12. Interference fringes of the LC lens when the optical axis is shifted to different positions. (**a–e**) multiple directions, (**f**) the total RMS aberration of the OAM lens. The red dot is the center of the lens.

Unlike the previous unidirectional movement, the applying voltage V_a and p of the eight electrodes remained, but the applying positions needed to be changed. Assuming the driving voltage at the top left corner position is V_{leftup}, V_{leftup} was listed in Table 5. The total root-mean-square (RMS) errors of the wavefront are then extracted and are shown in Figure 12f. It can be seen that OAM lenses have a small RMS error during multi-directional movement.

In summary, through the selection of various m and t, we have verified the feasibility of equidistant OAM LC lenses, constant optical axis offset LC lenses, and arbitrary range OAM LC lenses. The theoretical movement amount was compared with the actual OAM, and the consistency of the two datasets verified the effectiveness of the simulation for the actual voltage application situation. The lens can be moved to infinity, which is the horizontal stripe. The maximum optical power is ±6D, and the minimum focal length is 16.7 cm.

Table 5. The driving voltage and parameter of optical axis multi-directional movement.

m (V·mm^{-2})	t (V·mm^{-1})	s (mm)	V_{leftup} (V)	p (V)	Azimuth Angle
1.0	0.60	0.30	1.90	2.20	45° (left up)
1.0	0.60	0.30	4.10	2.20	45° (left down)
1.0	0.60	0.30	4.10	0	0°
1.0	0.60	0.30	4.10	2.20	45° (right up)
1.0	0.60	0.30	1.90	2.20	45° (right down)

3.4. *Phase Delay Response Time*

When no phase difference is generated, the eight electrodes of the OAM lens can be applied with the same voltages at 0 V and 4.1 V, but under the condition of 0 V, the response time of the OAM lens is too long to take 2–3 s, which will limit its application. However, when applying the voltage at 4.1 V, the response time of the OAM lens is significantly shortened for the LC, which is driven by pre-voltage. In this paper, the thickness of the LC is 30 μm, and response times can reach 0.635 s in Figure 13, which is regularly 2–3 s [21].

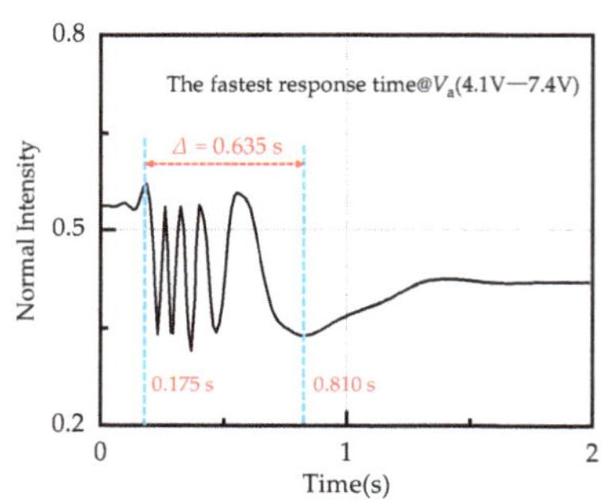

Figure 13. Response time of LC from initial state to maximum working state. The blue lines represent the start and end of the response time, respectively. The response time is 0.635 s.

The response time is measured based on the intensity changes during the variations of the fringes. The response time of the nematic LC lens is expressed as [41]

$$\tau_r = \frac{\gamma d^2}{\Delta\varepsilon(V^2 - V_{th}^2)} \quad (13)$$

γ: the visco-elastic coefficient, d: the thickness of the LC layer, V: driving voltage, V_{th}: threshold voltage. In practical applications, it is possible to use a low-viscosity LC material or reduce the thickness of the LC layer to increase the response speed. Moreover, the overdrive voltage method and undershoot effect can be used to reduce the response time [42].

4. Conclusions and Discussion

In conclusion, an LC lens with a movable optical axis is proposed and experimentally demonstrated firstly without mechanical movement. It can also function as a 3D wedge with a 0.164° steering angle. Experimentally, the vibration of the optical axis ranges from 0.164°, and the movement of the optical axis can reach infinity and remain unchanged during the movement, but there is a trade-off between movement and dioptre power of the OAM LC lens. In addition to the extremely shortened response time, it can be used for OIS. The comparison is listed in Table 6.

Table 6. Works compared with other typical papers.

LC Devices	Beam Deflectors			OAM Lens			Response Time
	Angle	Dimension	Thickness	Movement	Dimension	Aberration	
Hands [26]	0.024°	3D	50 μm	—	—	—	2–3 s
Ye [13]	0.1°	2D	130 μm	—	—	—	—
J.F.Algorri [35]	0.008°	2D	50 μm	—	—	—	—
Feng [37]	—	—	50 μm	1.0 mm	1D	< 0.1 λ	2–3 s
This paper	0.164°	3D	30 μm	0~∞ (firstly)	2D	< 0.1 λ	0.635–3 s

For an LC with normal dispersion, the refractive index difference Δn is a function of wavelength.

$$\Delta n = n_e - n_o = a + \frac{b}{\lambda^2} + \frac{c}{\lambda^4} \tag{14}$$

a, b, and c are Cauchy dispersion coefficients related to the liquid crystal material. The maximum dioptre power (OP_{max}) of the LC lens can be expressed as

$$OP_{max} = \frac{2d}{r^2}(a + \frac{b}{\lambda^2} + \frac{c}{\lambda^4}) \tag{15}$$

It can be seen that even if the thickness of the LC layer, the size of the aperture, and the birefringence difference of the LC lens are determined, its dioptre power will still be affected by wavelength. When polychromatic light passes through the LC lens and becomes a converging/diverging spherical wave, it will still be affected by the chromatic dispersion of the LC layer. This causes the focal point of different wavelengths to be separated in different image planes, resulting in the difference in retardation at different wavelengths of the liquid crystal lens. This effect will become more severe as the thickness of the liquid crystal layer and the difference in the birefringence of the liquid crystal lens increase.

Supplementary Materials: The following supporting information can be downloaded at https://www.mdpi.com/article/10.3390/cryst14100843/s1; Video S1: OAM LC lens.

Author Contributions: Conceptualization, Q.W. and H.Z.; methodology, Q.W. and H.Z.; validation, Q.W., H.Z., D.J. and T.L.; investigation, Q.W.; data curation, Q.W., H.Z., D.J. and T.L.; writing—original draft preparation, Q.W.; writing—review and editing, H.Z., D.J. and T.L.; supervision, D.J. and T.L.; project administration, H.Z., D.J. and T.L. All authors have read and agreed to the published version of the manuscript.

Funding: This work was funded by the National Natural Science Foundation of China (42175097).

Institutional Review Board Statement: Not applicable.

Informed Consent Statement: Not applicable.

Data Availability Statement: Data are provided within the manuscript or Supplementary Information files. The author confirms that all data generated or analysed during this study are included in this published article. Furthermore, primary and secondary sources and data supporting the findings of this study were all publicly available at the time of submission.

Conflicts of Interest: The authors declare no conflicts of interest.

References

1. Sun, M.; Feng, Y.; Wang, Y.; Huang, W.; Su, S. Design, Analysis and Experiment of a Bridge-Type Piezoelectric Actuator for Infrared Image Stabilization. *Micromachines* **2021**, *12*, 1197. [CrossRef] [PubMed]
2. Bleier, B.J.; Yezer, B.A.; Freireich, B.J.; Anna, S.L.; Walker, L.M. Droplet-Based Characterization of Surfactant Efficacy in Colloidal Stabilization of Carbon Black in Nonpolar Solvents. *J. Colloid Interface Sci.* **2017**, *493*, 265–274. [CrossRef] [PubMed]
3. Karasikov, N.; Peled, G.; Yasinov, R.; Feinstein, A. *Piezo-Based, High Dynamic Range, Wide Bandwidth Steering System for Optical Applications*; Pham, T., Kolodny, M.A., Eds.; SPIE: Anaheim, CA, USA, 2017; p. 101901C.

4. Zheng, Y.; Fu, Y.G.; He, W.J.; Wang, J.K.; Zhang, L. Opto-Mechanical Structure Design of the Image Stabilization Collimator. *Key Eng. Mater.* **2013**, *552*, 79–84. [CrossRef]
5. Meng, Z.; Huang, W.; Zhang, L.; Zhou, X.; Zhao, K.; Pu, D.; Chen, L. Large Aperture and Defect-Free Liquid Crystal Planar Optics Enabled by High-Throughput Pulsed Polarization Patterning. *Opt. Express* **2023**, *31*, 30435. [CrossRef]
6. Yoshida, S.; Schmid, W.; Vo, N.; Calabrase, W.; Kisley, L. Computationally-Efficient Spatiotemporal Correlation Analysis Super-Resolves Anomalous Diffusion. *Opt. Express* **2021**, *29*, 7616. [CrossRef]
7. Sun, Z.; Yuan, Z.; Nikita, A.; Kwok, H.; Srivastava, A.K. Fast-Switchable, High Diffraction-Efficiency Ferroelectric Liquid Crystal Fibonacci Grating. *Opt. Express* **2021**, *29*, 13978. [CrossRef]
8. Cheng, K.; Li, Z.; Wu, J.; Hu, Z.-D.; Wang, J. Super-Resolution Imaging Based on Radially Polarized Beam Induced Superoscillation Using an All-Dielectric Metasurface. *Opt. Express* **2022**, *30*, 2780. [CrossRef]
9. Chen, H.; Yu, S. A Stereo Camera Tracking Algorithm Used in Glass-Free Stereoscopic Display System. *J. Comput. -Aided Des. Comput. Graph.* **2017**, *29*, 436–443.
10. Li, R.; Zhang, H.; Chu, F.; Wang, Q. Compact Integral Imaging 2D/3D Compatible Display Based on Liquid Crystal Micro-Lens Array. *Liq. Cryst.* **2022**, *49*, 512–522. [CrossRef]
11. Dong, J.; Li, Z.; Liu, X.; Zhong, W.; Wang, G.; Liu, Q.; Song, X. High-Speed Real 3D Scene Acquisition and 3D Holographic Reconstruction System Based on Ultrafast Optical Axial Scanning. *Opt. Express* **2023**, *31*, 21721. [CrossRef]
12. Juodkazis, S.; Shikata, M.; Takahashi, T.; Matsuo, S.; Misawa, H. Fast Optical Switching by a Laser-Manipulated Microdroplet of Liquid Crystal. *Appl. Phys. Lett.* **1999**, *74*, 3627–3629. [CrossRef]
13. Kawamura, M.; Ye, M.; Sato, S. Optical Tweezers System by Using a Liquid Crystal Optical Device. *Mol. Cryst. Liq. Cryst.* **2007**, *478*, 135–891. [CrossRef]
14. Wu, Q.; Zhang, H.; Jia, D.; Liu, T. Recent Development of Tunable Optical Devices Based on Liquid. *Molecules* **2022**, *27*, 8025. [CrossRef] [PubMed]
15. Maksymova, I.; Greiner, P.; Wiesmeier, J.; Darrer, F.M.; Druml, N. A MEMS Mirror Driver ASIC for Beam-Steering in Scanning MEMS-Based LiDAR. In Proceedings of the Laser Beam Shaping XIX, San Diego, CA, USA, 11–12 August 2019; Dudley, A., Laskin, A.V., Eds.; SPIE: San Diego, CA, USA, 2019; p. 11.
16. Galaktionov, I.; Kudryashov, A.; Sheldakova, J.; Nikitin, A. Laser Beam Focusing through the Dense Multiple Scattering Suspension Using Bimorph Mirror. In Proceedings of the Adaptive Optics and Wavefront Control for Biological Systems V, San Diego, CA, USA, 3–4 February 2019; Bifano, T.G., Gigan, S., Ji, N., Eds.; SPIE: San Francisco, CA, USA, 2019; p. 44.
17. Wlodarczyk, K.L.; Bryce, E.; Schwartz, N.; Strachan, M.; Hutson, D.; Maier, R.R.J.; Atkinson, D.; Beard, S.; Baillie, T.; Parr-Burman, P.; et al. Scalable Stacked Array Piezoelectric Deformable Mirror for Astronomy and Laser Processing Applications. *Rev. Sci. Instrum.* **2014**, *85*, 024502. [CrossRef]
18. Verpoort, S.; Wittrock, U. Unimorph Deformable Mirror for Telescopes and Laser Applications. In Proceedings of the International Conference on Space Optics—ICSO 2010, Rhodes Island, Greece, 4–8 October 2010; Armandillo, E., Cugny, B., Karafolas, N., Eds.; SPIE: Rhodes Island, Greece, 2019; p. 527.
19. Salter, P.S.; Booth, M.J. Adaptive Optics in Laser Processing. *Light Sci. Appl.* **2019**, *8*, 10. [CrossRef]
20. Nose, T.; Sato, S. A Liquid Crystal Microlens Obtained with a Non-Uniform Electric Field. *Liq. Cryst.* **1989**, *5*, 1425–1433. [CrossRef]
21. Morris, R.; Jones, C.; Nagaraj, M. Liquid Crystal Devices for Beam Steering Applications. *Micromachines* **2021**, *12*, 247. [CrossRef]
22. He, Z.; Gou, F.; Chen, R.; Yin, K.; Zhan, T.; Wu, S.-T. Liquid Crystal Beam Steering Devices: Principles, Recent Advances, and Future Developments. *Crystals* **2019**, *9*, 292. [CrossRef]
23. Sato, S. Applications of Liquid Crystals to Variable-Focusing Lenses. *Opt. Rev.* **1999**, *6*, 471–485. [CrossRef]
24. Reshetnyak, V.Y.; Sova, O.; Wang, Y.-J.; Lin, Y.-H. Modeling Liquid Crystal Lenses. In Proceedings of the Emerging Liquid Crystal Technologies XV, San Diego, CA, USA, 3–5 February 2020; Chien, L.-C., Broer, D.J., Eds.; SPIE: San Francisco, CA, USA, 2020; p. 3.
25. Love, G.D.; Major, J.V.; Purvis, A. Liquid-Crystal Prisms for Tip-Tilt Adaptive Optics. *Opt. Lett.* **1994**, *19*, 1170. [CrossRef]
26. Hands, P.J.W.; Tatarkova, S.A.; Kirby, A.K.; Love, G.D. Modal Liquid Crystal Devices in Optical Tweezing: 3D Control and Oscillating Potential Wells. *Opt. Express* **2006**, *14*, 4525. [CrossRef] [PubMed]
27. Kotova, S.P.; Patlan, V.V.; Samagin, S.A. Tunable Liquid-Crystal Focusing Device. 2. Experiment. *Quantum Electron.* **2011**, *41*, 65–70. [CrossRef]
28. Xu, L.; Zhang, Y.; Liu, Z.; Ye, M. Driving Method for Adjustable Liquid Crystal Optical Wedge with Four Electrodes. *Acta Opt. Sin.* **2022**, *42*, 1323001. [CrossRef]
29. Ye, M.; Sato, S. Liquid Crystal Lens with Focus Movable along and off Axis. *Opt. Commun.* **2003**, *225*, 277–280. [CrossRef]
30. Ye, M.; Wang, B.; Sato, S. Liquid Crystal Lens with Focus Movable in Focal Plane. *Opt. Commun.* **2006**, *259*, 710–722. [CrossRef]
31. Ye, M.; Wang, B.; Uchida, M.; Yanase, S.; Takahashi, S.; Yamaguchi, M.; Sato, S. Low-Voltage-Driving Liquid Crystal Lens. *Jpn. J. Appl. Phys.* **2010**, *49*, 100204. [CrossRef]
32. Bennis, N.; Jankowski, T.; Strzezysz, O.; Pakuła, A.; Zografopoulos, D.C.; Perkowski, P.; Sánchez-Pena, J.M.; López-Higuera, J.M.; Algorri, J.F. A High Birefringence Liquid Crystal for Lenses with Large Aperture. *Sci. Rep.* **2022**, *12*, 14603. [CrossRef]
33. Bennis, N.; Jankowski, T.; Morawiak, P.; Spadlo, A.; Zografopoulos, D.C.; Sánchez-Pena, J.M.; López-Higuera, J.M.; Algorri, J.F. Aspherical Liquid Crystal Lenses Based on a Variable Transmission Electrode. *Opt. Express* **2022**, *30*, 12237. [CrossRef]
34. Algorri, J.F.; Love, G.D.; Urruchi, V. Modal Liquid Crystal Array of Optical Elements. *Opt. Express* **2013**, *21*, 24809. [CrossRef]

35. Algorri, J.F.; Morawiak, P.; Zografopoulos, D.C.; Bennis, N.; Spadlo, A.; Rodríguez-Cobo, L.; Jaroszewicz, L.R.; Sánchez-Pena, J.M.; López-Higuera, J.M. Multifunctional Light Beam Control Device by Stimuli-Responsive Liquid Crystal Micro-Grating Structures. *Sci. Rep.* **2020**, *10*, 13806. [CrossRef]
36. Algorri, J.F.; Morawiak, P.; Bennis, N.; Zografopoulos, D.C.; Urruchi, V.; Rodríguez-Cobo, L.; Jaroszewicz, L.R.; Sánchez-Pena, J.M.; López-Higuera, J.M. Positive-Negative Tunable Liquid Crystal Lenses Based on a Microstructured Transmission Line. *Sci. Rep.* **2020**, *10*, 10153. [CrossRef] [PubMed]
37. Feng, W.; Liu, Z.; Liu, H.; Ye, M. Design of Tunable Liquid Crystal Lenses with a Parabolic Phase Profile. *Crystals* **2022**, *13*, 8. [CrossRef]
38. Cao, F.; Liu, Z.; Ye, M. Modulated Beam Deflection by Liquid Crystal Optical Wedge Arrays. *Acta Opt. Sin.* **2024**, *44*, 251–257. [CrossRef]
39. Elston, S.J. Optics and Nonlinear Optics of Liquid Crystals. *J. Mod. Opt.* **1994**, *41*, 1517–1518. [CrossRef]
40. Feng, W.; Liu, Z.; Ye, M. Positive-Negative Tunable Cylindrical Liquid Crystal Lenses. *Optik* **2022**, *266*, 169613. [CrossRef]
41. Xu, S.; Li, Y.; Liu, Y.; Sun, J.; Ren, H.; Wu, S.-T. Fast-Response Liquid Crystal Microlens. *Micromachines* **2014**, *5*, 300–324. [CrossRef]
42. Yang, D.-K.; Wu, S.-T. *Fundamentals of Liquid Crystal Devices*; John Wiley & Sons: Hoboken, NJ, USA, 2014. [CrossRef]

crystals

MDPI

Review

Review of Angular-Selective Windows with Guest–Host Liquid Crystals for Static Window Applications

Chan-Heon An and Seung-Won Oh *

Department of Electrical Information Communication Engineering, Kangwon National University, Samcheok, Gangwon 25913, Republic of Korea
* Correspondence: ohseungwon@kangwon.ac.kr

Abstract: This review focuses on the development and advancements in angular-selective smart windows, with particular emphasis on static windows utilizing guest–host liquid crystal (GHLC) systems. Angular-selective windows are designed to adjust their transmittance based on the angle of incident light, offering enhanced energy efficiency and visual comfort in both architectural and automotive applications. By leveraging the anisotropic absorption properties of dichroic dyes, GHLC-based windows can selectively block oblique sunlight while preserving clear visibility from normal viewing angles. Various liquid crystal (LC) alignment configurations, including vertically aligned, homogeneously aligned, hybrid aligned, uniformly lying helix, and twisted aligned LC cells, have been investigated to optimize light control for different installation angles, such as for automotive windshields and building windows. These advancements have demonstrated significant improvements in energy conservation and occupant comfort by reducing cooling demands and regulating sunlight penetration. This review summarizes key findings from recent studies, addresses the limitations of current technologies, and outlines potential future directions for further advancements in smart window technology.

Keywords: liquid crystals; smart window; static window; angular dependence; absorption

1. Introduction

Windows play a critical role in modern architecture, balancing the need for natural light, ventilation, and thermal regulation. Smart windows, which can dynamically or statically modulate the amount of sunlight and solar heat entering a space, have emerged as a key technology for improving energy efficiency while maintaining visual comfort [1–8]. In the context of escalating concerns about global warming and energy consumption, the development of energy-efficient solutions in buildings has become paramount. Traditional solutions, such as tinted windows, often fall short because they cannot adjust based on the angle of sunlight, leading to issues like glare and heat buildup. To address this, both dynamic and static smart windows have been developed, each offering unique advantages depending on the application.

Dynamic smart windows utilize systems that incorporate chromic materials, including electrochromic, photochromic, and thermochromic technologies. These windows modify their transmittance in reaction to external stimuli, such as variations in light, heat, or applied voltage [9–14]. While these dynamic windows offer flexibility in controlling light and heat, they generally absorb light isotropically, blocking sunlight from all directions equally. This uniform control can limit visibility and does not account for the angle of incoming light, which can be problematic in situations requiring selective sunlight control.

In contrast, angular-selective smart windows, often referred to as static windows, offer a more targeted approach [15–18]. These windows, particularly those utilizing guest–host liquid crystal (GHLC) systems [19–21], are designed to selectively block sunlight from specific angles, such as oblique incident light, while maintaining a clear view from direct, normal angles. This angular dependence is achieved through the anisotropic absorption properties of dichroic dyes, making static windows particularly effective in managing glare and heat from specific directions. For instance, static smart windows can reduce intense solar heat and glare during peak sunlight hours without darkening the entire view, thus minimizing the need for artificial lighting or excessive cooling. This capability is especially relevant for zero-energy buildings, where maximizing energy efficiency is essential to meet stringent sustainability goals and reduce reliance on external energy sources.

The distinction between static and dynamic smart windows is crucial as it highlights different approaches to optimizing energy efficiency and visual comfort. While dynamic windows excel in providing adjustable transparency in response to changing external conditions, static windows offer more refined control based on the angle of sunlight. This makes static windows ideal for applications such as automotive windshields, where reducing glare from oblique angles is critical for safety, or in buildings with large window surfaces, where selective light transmission can significantly reduce cooling requirements while maintaining indoor comfort. In a broader context, integrating angular-selective smart windows into architectural design can contribute significantly to global efforts aimed at combating climate change by reducing energy consumption and promoting sustainable living.

In this review, we explore the advancements in angular-selective smart windows, with a focus on GHLC systems. We examine various design strategies, performance metrics, and energy-saving potential of these windows, highlighting their distinct advantages over traditional dynamic windows. We also explore existing technologies and future research strategies that have the potential to enhance energy efficiency and improve occupant comfort, thereby contributing to the creation of more sustainable built environments.

2. Static (Angular-Selective) Windows

Most conventional studies on light-absorbing windows have primarily focused on normal incidence, where light enters the window directly at a perpendicular angle. However, this approach does not adequately reflect real-world scenarios. In real-world applications, sunlight usually reaches windows at oblique angles due to varying solar elevations throughout the day and across different seasons [19]. According to the data of the sun path in Korea during the summer and winter solstices [22], in summer, the solar elevation angle peaks around midday, causing sunlight to hit windows from a higher angle. Conversely, during winter, the solar elevation is lower, resulting in sunlight entering at shallower angles. These seasonal variations not only influence the intensity of sunlight but also the quality of light that enters the building, affecting both the esthetic and functional aspects of interior spaces.

To achieve optimal energy efficiency, controlling the transmission of sunlight based on these varying angles is crucial [19,21]. During the summer, absorbing more sunlight at higher angles can help reduce the heat entering the building, thus lowering cooling energy demands. This is particularly important in regions with hot climates, where excessive heat can lead to discomfort and increased reliance on air conditioning systems, significantly driving up energy consumption. On the other hand, in the winter, transmitting more light at lower angles can reduce heating and artificial lighting requirements (Figure 1a). By allowing more sunlight to enter, buildings can harness natural light, which is beneficial not only for energy savings but also for creating a more pleasant indoor environment. This ability to adapt to both seasonal and daily changes in solar elevation makes angular-selective windows an effective solution for energy saving in both commercial and residential settings.

Beyond their application in buildings, static windows are also highly beneficial in automotive contexts (Figure 1b) [20]. Many drivers use tinted windows, sunglasses, or sun visors to reduce glare and block sunlight. However, safety regulations in countries such as the United States and across Europe often limit windshield tinting to a transmittance of 50–70%, which can be insufficient for blocking sunlight and glare during driving. This regulatory limitation highlights the need for innovative solutions that prioritize both safety and comfort. Traditional solutions like sunglasses and sun visors also come with their own drawbacks, including frequent adjustments and restricted visibility. Moreover, these conventional methods may not provide adequate protection against UV radiation, which can have harmful effects over prolonged exposure.

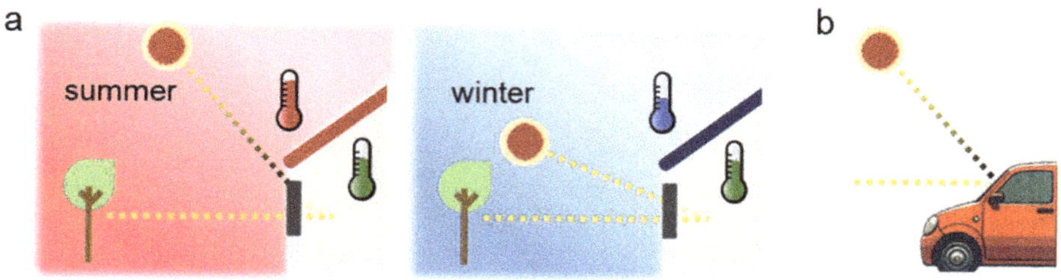

Figure 1. (**a**) Schematic representation of the energy-saving window's performance in summer and winter. (**b**) Schematic illustration of the windshield for automotive.

Static windows utilizing GHLC cells offer a more efficient alternative. These windows can block oblique sunlight while maintaining clear visibility through normal angles, making them ideal for both buildings and vehicles. For example, when the solar elevation angle exceeds 75°—common during certain times of the day and year—GHLC windows effectively block direct sunlight without obstructing the driver's view. This feature is especially crucial during dawn and dusk when the sun is low on the horizon, as these are often the times when glare is most problematic for drivers. This angular selectivity offers a significant advantage over traditional static solutions, such as tinted glass, by dynamically adjusting the transmittance based on sunlight's incident angle. This results in improved comfort and safety for drivers.

Overall, static windows provide a necessary solution for improving both energy efficiency and occupant comfort across various environments. As we confront global challenges like climate change and the urgent need for sustainable living practices, the adoption of such innovative window technologies becomes increasingly critical. By selectively absorbing or transmitting light based on the angle of incidence, these windows offer a flexible and efficient method for managing solar radiation, contributing to reduced energy consumption and enhanced user experience in both architectural and automotive applications. Their integration into building designs and vehicle manufacturing processes can significantly contribute to the broader goals of energy conservation and environmental sustainability.

3. Principles of Designing Static Windows Using GHLC

GHLC devices, which consist of a host liquid crystal (LC) and guest dichroic dye, have gained significant attention due to their high transmittance and polarizer-free structure [23,24]. This section outlines the fundamental principles that underlie the design and operation of GHLC static windows, highlighting their unique properties and potential applications. The unique optical characteristics of GHLC devices stem from the dichroism of incorporated dye molecules. The transmittance of a homogenously aligned (HA) dye-doped LC cell can be described using the following mathematical expression [25–27]:

$$T_{\parallel} = T_0 e^{-\alpha_{\parallel} c d} \tag{1a}$$

$$T_{\perp} = T_0 e^{-\alpha_{\perp} c d} \tag{1b}$$

where $T_{\parallel}$ and $T_{\perp}$ refer to the transmittance of the GHLC cell when the polarization is aligned parallel and perpendicular to the dye molecules' absorption axis, respectively. $\alpha_{\parallel}$ and $\alpha_{\perp}$ represent the absorption coefficients of dye-doped LCs for polarization parallel and perpendicular to the absorption axis of the dye molecules, respectively. These dye molecules display a significant absorption of light polarized along their absorption axis, while exhibiting minimal absorption for light polarized orthogonally to this axis (Figure 2a). T_0 represents the transmittance of an LC cell without dye, while c and d denote the concentration of the dye gap and cell gap, respectively.

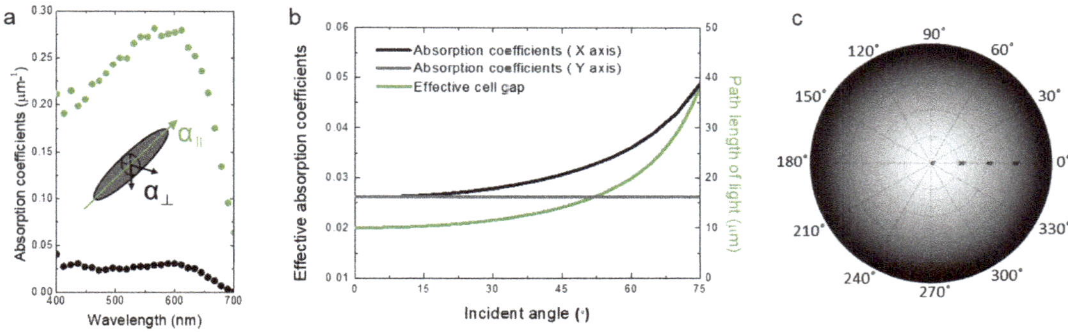

Figure 2. (**a**) Absorption coefficients of dichroic dye mixture X12 across different wavelengths. (**b**) Variation in effective absorption coefficients and optical path length as a function of increasing incident angles. (**c**) Contour plot illustrating the transmittance.

When designing GHLC static windows, it is essential to consider the angle of incident light as a variable. Both the absorption coefficient and effective cell gap are influenced by this angle, increasing as the angle of incidence rises. The effective absorption coefficient can be calculated as follows:

$$\frac{1}{\alpha_{eff}^2(\theta)} = \frac{sin^2\theta}{\alpha_{\parallel}^2} + \frac{cos^2\theta}{\alpha_{\perp}^2} \tag{2}$$

where θ is the angle of the incident light. As the incident angle increases, the tilt angle of the dye molecules also increases, leading to enhanced absorption for light polarized parallel to the absorption axis (Figure 2b).

Furthermore, the effective cell gap is affected by the angle of incident of the light and can be computed using basic trigonometric functions:

$$d_{eff}\,(\theta) = d_0/cos\ \theta \tag{3}$$

Here, d_0 is the cell gap perpendicular to the substrates, and d_{eff} indicates the effective cell gap as a function of the incident light angle. This demonstrates that as the angle of incidence increases, the effective cell gap also increases, which correlates with a decrease in transmittance (Figure 2b).

Figure 2c illustrates the transmittance contour of the vertically aligned (VA) GHLC cell. The VA cell shows a steady decrease in transmittance in all directions as the angle of incidence increases. GHLC films, in contrast, have shown promise for various applications, including privacy displays and reducing reflections on automotive windshields. While VA cells have traditionally been favored for their simplicity, it is crucial to recognize that they may compromise viewing angles for end-users.

The integration of GHLC technology into static windows offers a distinct advantage: it enables the efficient management of light transmission while maintaining high visual clarity. By leveraging the angular-selective absorption properties of GHLC devices, designers can create windows that not only optimize energy efficiency but also enhance user comfort by minimizing glare and maintaining clear visibility. In summary, the design principles of GHLC static windows revolve around the exploitation of dichroic properties, the consideration of angular dependence, and the optimization of performance characteristics.

This innovative approach paves the way for developing advanced window solutions that address the needs of various applications while promoting energy conservation.

4. Effects of Different LC Alignments on Static Windows

Figure 3a is a 3D schematic representation of the GHLC device in the VA state. The incident angle of sunlight increases relative to the normal view (z-direction). The GHLC devices exhibit different transmittance depending on the liquid crystal alignment state (Figure 3b). The transmittances of LC cells in different alignment states—VA, HA, uniformly lying helix (ULH), and twist-aligned (TA)—can be expressed using Equation (4) [28]. We used unpolarized incident light for the calculations in the equation. Since the effective absorption coefficients differ in the x- and y-directions, we separated the incident light into these two directions. We then calculated the transmittance for each direction individually and superimposed the results to obtain the final transmittance. Each alignment state affects the behavior of dye molecules and their interaction with polarized light in unique ways, influencing the transmittance characteristics. The transmittances of states can be expressed, respectively, as

$$T_{\mathrm{VA}} = T_0 exp(-\alpha_{\perp} cd) \tag{4a}$$

$$T_{\mathrm{HA}} = T_0 \left[exp\left(-\alpha_{\parallel} cd\right) + exp(-\alpha_{\perp} cd) \right] / 2 \tag{4b}$$

$$T_{\mathrm{ULH}} = T_0 \left[exp\left(-\left(\alpha_{\parallel} + \alpha_{\perp}\right) cd/2\right) + exp(-\alpha_{\perp} cd) \right] / 2 \tag{4c}$$

$$T_{\mathrm{TA}} = T_0 exp\left[-\left(\alpha_{\parallel} + \alpha_{\perp}\right) cd/2\right] \tag{4d}$$

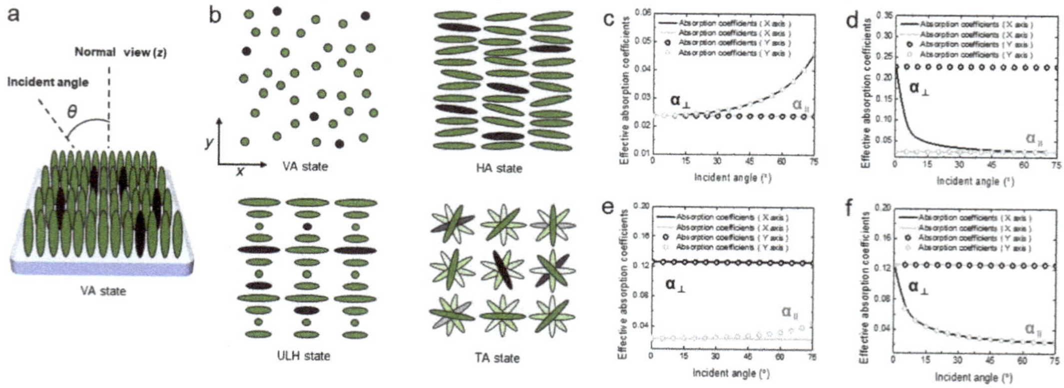

Figure 3. (**a**) 3D schematic of the GHLC cell in the VA state. (**b**) Top view of the GHLC cells in different LC states. (**c–f**) Effective absorption coefficients in the VA, HA, ULH, and TA states.

In the VA state, dye molecules are aligned perpendicular to the substrate, making the window transparent at normal incidence. In the HA state, they selectively absorb light polarized along the rubbing direction. In the ULH state, LCs and dyes are twisted, absorbing light thar is polarized perpendicular to the helical axis and parallel to the dichroic dye's principal axis. In the TA state, dye molecules absorb light regardless of their polarization.

For static windows, the aim is to maintain transparency at normal viewing angles and opacity at oblique angles. Therefore, the difference in transmittance between normal and oblique views is crucial for performance. The absorption coefficients for different LC states were calculated (Figure 3c–f), showing how molecular structure and absorption vary with the incidence angle in the x-, y-, and z-directions.

In the VA state, as the incidence angle increases, the molecules gradually shift from vertical to horizontal (HA state), leading to increased absorption (Figure 3c). In the HA state, the absorption coefficient remains unchanged in the x-direction but decreases in the y-direction as the angle increases (Figure 3d). In the ULH state, as the angle increases in the x-direction, the molecular structure shifts towards the TA state, resulting in increased absorption. In the y-direction, the structure remains constant, with no significant change in absorption (Figure 3e). In the TA state, the molecular alignment shifts to a helical structure at higher angles, reducing absorption in both the x- and y-directions (Figure 3f).

These variations in absorption coefficients based on LC alignment and incident angles are key to optimizing static window performance, allowing for control over light absorption while maintaining transparency at specific viewing angles.

5. Static Windows for Energy-Saving Building

In order to assess the angular transmittance of GHLC cells featuring various configurations, we utilized the "TechWiz LCD 2D" software, which employs the Berreman matrix method for light transmission calculations. Static windows do not rely on polarizers and are designed to block unpolarized sunlight; the transmittance of the GHLC cells was calculated by an unpolarized white light source. The primary objective was to determine the optimal dye concentration and cell gap by analyzing the transmittance variations in both the x- and y-directions. Using the algorithm for optimization [28], we set the combination of LC and dye that showed the highest transmittance difference. The most effective configuration included ML-1309 LC material (nematic phase at room temperature), characterized by an optical birefringence (Δn) of 0.1096 and a dielectric anisotropy ($\Delta\varepsilon$) of −3.9 from Merck mixed with the black dye X12 from BASF, as shown in Figure 2a.

We confirmed the absorption coefficients using this specific LC-dye mixture. The process involved a horizontally aligned cell that allowed for the evaluation of the absorption coefficient for horizontally polarized light, assessed both parallel and perpendicular to the absorption axis. The X12 dye exhibited absorption coefficients of $\alpha_{\perp}$ = 0.024 μm^{-1} and $\alpha_{\parallel}$ = 0.227 μm^{-1} within the wavelength range of 400 to 700 nm. For the VA and HA cells, we maintained a cell gap of 10 μm with a dye concentration of 1 wt%. In the case of the ULH cell, we incorporated a chiral dopant (S811, HTP ~11) and established a pitch of 8.75 μm, noting that variations in pitch did not significantly influence transmittance. We selected a cell gap of 5.5 μm and a dye concentration of 0.6 wt%. For the TA cell, the same chiral dopant was utilized, and a pitch of 1000 nm was established to reduce light reflection. This cell configuration also had a cell gap of 10 μm, with a dye concentration of 2 wt%.

This section provides a detailed analysis of the transmittance properties of various GHLC cells—VA, HA, ULH, and TA—across different incident angles. As presented in Figure 4a,b, software simulations revealed substantial differences in transmittance based on cell structure and incident angle. For instance, the VA cell exhibited the largest transmittance difference between normal and oblique angles, especially in the z- and y-directions at 75°, showing transmittance values of 60.3% and 18.4%, respectively. This resulted in a discrepancy of 41.9%, making it the most effective at controlling sunlight. By comparison, the TA cell showed minimal transmittance differences, with values of 3.73% and 2.48%, respectively, resulting in a small discrepancy of 1.26%.

In the x-direction, both VA and TA cells demonstrated similar transmittance behaviors, limiting the field of view for occupants. However, the HA and ULH cells showed significantly lower transmittance differences of 8.85% and 12.5% in the z- and x-directions, ensuring a broader and more stable viewing angle, particularly in the horizontal plane (x-direction) within a ± 60° range. This feature of the HA and ULH cells provides a nearly uniform transmittance, maintaining the visual comfort of building occupants while effectively reducing glare.

Figure 4c–f further highlights the transmittance contours of these cells, with the VA cell showing a consistent decline in transmittance across all directions as the incident angle increases. In contrast, the HA and ULH cells exhibit more controlled transmittance variation, particularly in the x-direction, making them ideal candidates for applications requiring wide viewing angles and controlled light absorption. The TA cell, with its uniform transmittance across all angles, can be useful in applications where uniform light blocking is desired.

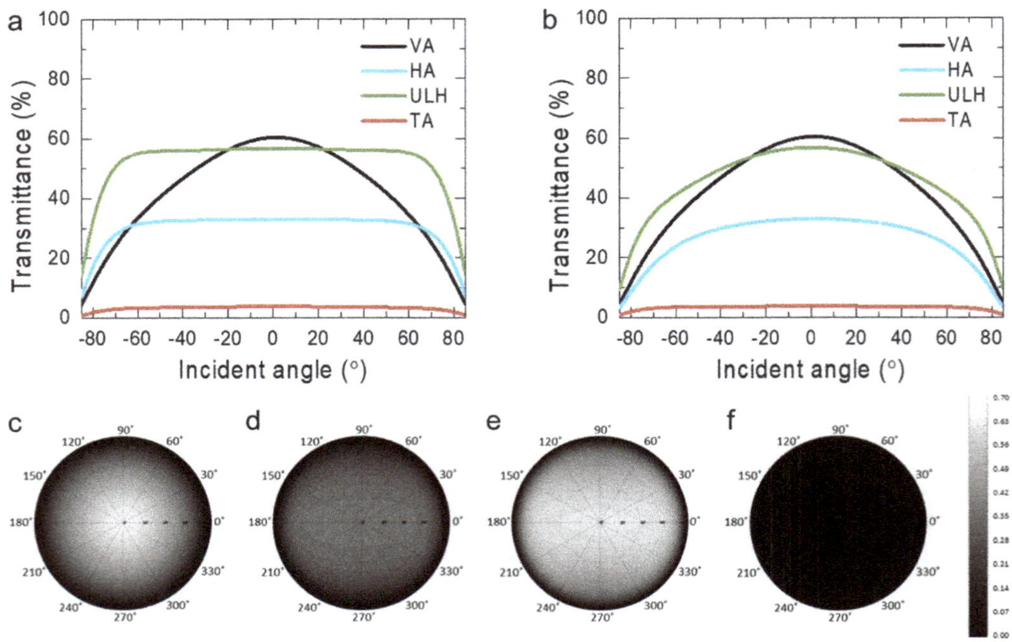

Figure 4. Transmittance of VA, HA, ULH, and TA cells in (**a**) $x-$ and (**b**) $y-$directions as a function of increasing incident angles. (**c–f**) Transmittance contours of VA, HA, ULH, and TA cells.

This comparison underscores that the ULH cell may serve as a promising candidate for building applications due to its balance between light control and a wide viewing angle, while the TA cell's uniform transmittance makes it suitable for cases where blocking all incident light is required.

To gain a deeper understanding of how varying incident angles affect light absorption in VA, HA, ULH, and TA cells, we carried out a detailed series of simulations. These simulations were performed at incident angles ranging from 0° to 75°, with increments of 15° for each trial (refer to Figure 5 for visual representation). As the angles of incidence increased, distinct behaviors emerged for each cell type. In the case of the VA cell, the structure appeared progressively darker as the incident angle increased, showing uniform behavior regardless of the viewing direction. This consistent darkening suggests that VA cells are highly sensitive to variations in incident angles. On the other hand, the HA and ULH cells, particularly in the x-direction, showed significantly lower light absorption, maintaining their brightness even at incident angles as steep as 60°. This performance highlights the robustness of these cells in managing light absorption across a wide range of viewing angles.

Although the static window configuration we tested does not possess dynamic transmittance adjustment, it provides several practical benefits. Most notably, it can help reduce energy consumption by limiting heat from sunlight, thus lowering air conditioning costs, while simultaneously enhancing the overall visual comfort for users. Additionally, our findings suggest that HA and ULH cells exhibit higher transmittance in the y-direction compared to the VA cell, revealing an area that warrants further investigation. One promising future direction is exploring methods to increase absorption in the y-direction without compromising the notable transmittance levels in the z- and x-directions, especially for applications requiring uniform light control. However, it is important to acknowledge the HA cell's limitations, particularly its relatively smaller transmittance differences, which may reduce its effectiveness in certain scenarios.

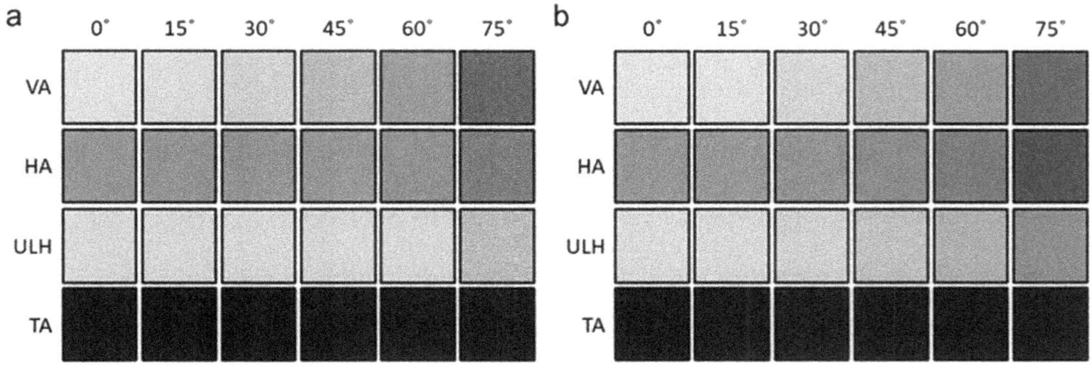

Figure 5. Calculated images of VA, HA, ULH, and TA cells in (**a**) x- and (**b**) y-directions as a function of increasing incident angles.

6. Static Windows for Automotive Applications

In automotive applications, drivers often rely on sunglasses or sun visors to reduce glare and block direct sunlight, as conventional tinted windows struggle to balance visibility with sunlight control [20]. Unlike static windows in energy-efficient buildings, which are optimized for maximum transmittance at normal incidence, automotive windshields present a unique challenge due to their installation angle. Traditional window designs, while effective for flat surfaces, face limitations when applied to windshields because the angle of installation affects the sunlight incidence throughout the day. This can lead to suboptimal sunlight control and reduced visibility in vehicles. Therefore, a new design approach is required—one that considers the windshield's installation angle to ensure clear visibility for drivers while efficiently blocking sunlight and reducing glare.

To evaluate the angular transmittance of GHLC cells for automotive applications, we used the software "TechWiz LCD 2D", maintaining consistent LC and dye material properties. For the tilted LC cell, we set a cell gap of 5.5 μm and a dye concentration of 1 wt%. The hybrid aligned nematic (HAN) cell also had a cell gap of 10 μm and a dye concentration of 0.3 wt%.

Figure 6a illustrates the tilted LC cell's schematic for the proposed installation angle, which typically ranges from 40° to 50°. In this study, we set the installation angle at 45°. Figure 6b shows the calculated transmittance of the GHLC cell with a pre-tilt angle of 55°. The reason the maximum transmittance angle of the LC does not exactly match the pre-tilt angle and installation angle is that the pre-tilt angle affects both the absorption coefficient and the path length. The highest normal-view transmittance for the tilted LC cell was 70.8% at a pre-tilt angle of 55°. In contrast, the transmittance in the oblique view (incident angle of 75°) was 50.8%, resulting in a 20% difference. Figure 6c presents the transmittance contours. The normal-view transmittance for automotive windshields was calculated at 70%, but this value can be adjusted for different installation angles by modifying the dye concentration and cell gap.

Figure 6d shows the HAN cell's schematic for the proposed installation angle. The HAN cell can adjust the highest transmittance angle by controlling the dye concentration or cell gap. Figure 6e presents the transmittance for the GHLC cell at a 45° installation angle. The highest normal-view transmittance for the HAN cell was also 69.7%, while the transmittance in the oblique view (75°) was 62.6%, again with a 7.1% difference. The transmittance contours are shown in Figure 6f. Similar to the tilted LC cell, the normal-view transmittance for automotive windshields was calculated to be 70%, but the transmittance can be tuned by adjusting the installation angle, dye concentration, and cell gap.

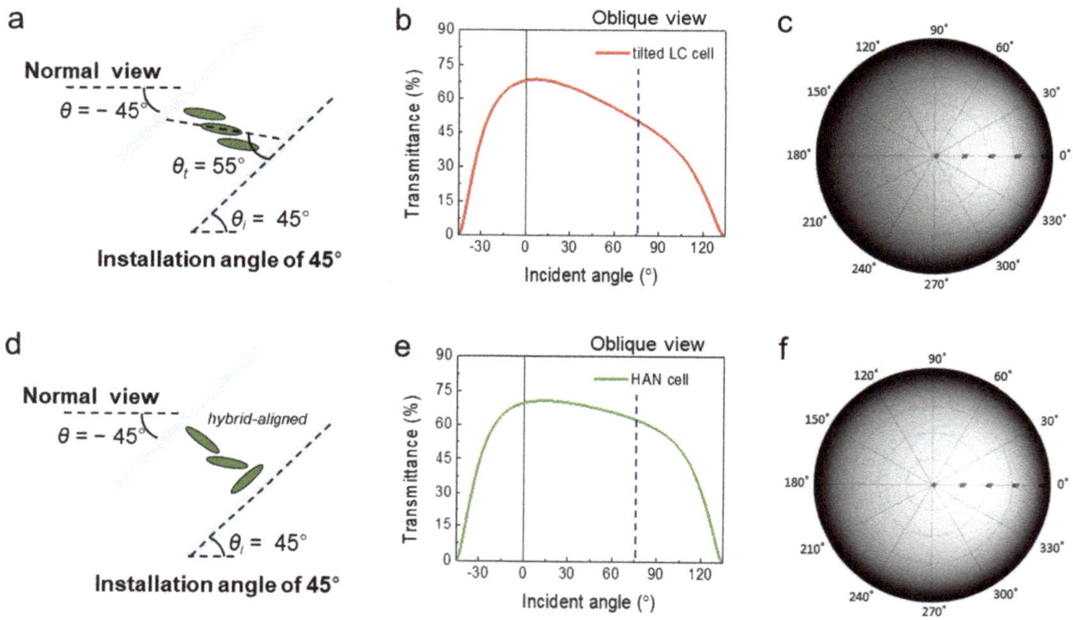

Figure 6. Tilted LC cell. (**a**) Schematic representation; (**b**) transmittance profile; (**c**) transmittance contour. HAN LC cell. (**d**) Schematic representation; (**e**) transmittance profile; (**f**) transmittance contour.

7. Conclusions and Future Perspectives

In conclusion, this study demonstrated the potential of various LC modes for static window applications, applicable to both energy-saving buildings and automotive environments. For building windows, we investigated VA, HA, ULH, and TA cells. Among these, the VA cell exhibited the largest transmittance difference between normal and oblique angles, making it particularly effective for sunlight control. In contrast, while the ULH cell showed a slightly smaller transmittance difference in the *y*-direction, it maintained stable transmittance in the *x*-direction, providing a wider viewing angle, which is advantageous for practical implementation.

In automotive applications, we addressed the limitations of traditional designs that focus on maximizing transmittance at a normal incidence—a strategy less effective for windshields due to their installation angle. To overcome this, we utilized tilted LC and HAN cells. The HAN cell, while easier to fabricate, exhibited relatively smaller transmittance differences, whereas the tilted LC cell provided more significant variation but required a more complex fabrication process. This trade-off between ease of manufacturing and performance highlights the need for specific design choices based on the intended application.

In addition to building energy savings and automotive applications, the angular-selective light absorption properties explored in this study have potential applications in fields where control over incident light angles is critical. One promising area is greenhouse agriculture, where managing the sunlight intensity and spectrum can significantly affect plant growth. Static windows with angular-selective properties can optimize natural light exposure throughout the day, reducing the need for artificial lighting and improving energy efficiency. Another possible application is in personal protective equipment, such as visors or eyewear, where selectively controlling light based on the viewing angle can enhance visibility and user comfort in varying light conditions.

Beyond the scope of visible light, it is important to note that near-infrared (NIR) light accounts for nearly 50% of the solar spectrum. However, the introduced GHLC devices in this study are limited in controlling NIR light due to the absorption range of the dyes used. To address this limitation, future designs should explore the integration of electrochromic devices that are capable of NIR control. However, electrochromic devices typically lack angular-selective properties, making them less suitable for static window applications. To overcome this challenge, research into reflection-based cholesteric LC devices may provide a viable solution. Such devices can selectively reflect or transmit NIR light depending on the angle of incidence, offering the potential for developing static windows that effectively control both visible and NIR light based on viewing angle.

Author Contributions: Writing—review and editing, C.-H.A.; supervision, S.-W.O. All authors have read and agreed to the published version of the manuscript.

Funding: This work was supported by the National Research Foundation of Korea (NRF) Grant funded by the Korean government (MSIP) (No. 2021R1I1A3052581).

Data Availability Statement: Data will be made available on reasonable request.

Conflicts of Interest: The authors declare no conflicts of interest.

References

1. Svensson, J.S.E.M.; Granqvist, C.G. Electrochromic coatings for "smart windows". *Sol. Energy Mater.* **1985**, *12*, 391–402. [CrossRef]
2. Baetens, R.; Jelle, B.P.; Gustavsen, A. Properties, requirements and possibilities of smart windows for dynamic daylight and solar energy control in buildings: A state-of-the-art review. *Sol. Energy Mater. Sol. Cells* **2010**, *94*, 87–105. [CrossRef]
3. Casini, M. Smart windows for energy efficiency of buildings. In Proceedings of the Second International Conference on Advances in Civil, Structural, and Environmental Engineering, Zurich, Switzerland, 26 October 2014; p. 273.
4. Ke, Y.; Chen, J.; Lin, G.; Wang, S.; Zhou, Y.; Yin, J.; Lee, P.S.; Long, Y. Smart windows: Electro-, thermo-, mechano-, photochromics, and beyond. *Adv. Energy Mater.* **2019**, *9*, 1902066. [CrossRef]
5. Madhuri, P.L.; Bhupathi, S.; Shuddhodana, S.; Judeh, Z.M.A.; Yang, S.H.; Long, Y.; Abdulhalim, I. Hybrid vanadium dioxide-liquid crystal tunable non-reciprocal scattering metamaterial smart window for visible and infrared radiation control. *Opt. Mater. Express* **2021**, *11*, 3023–3037. [CrossRef]
6. Xu, G.; Xia, H.; Chen, P.; She, W.; Zhang, H.; Ma, J.; Ruan, Q.; Zhang, W.; Sun, Z. Thermochromic Hydrogels with Dynamic Solar Modulation and Regulatable Critical Response Temperature for Energy-Saving SmartWindows. *Adv. Funct. Mater.* **2022**, *32*, 2109597. [CrossRef]
7. Pisharody, G.R.; Sahoo, P.; Rao, D.S.S.; Matte, H.S.S.R.; Sikar, D.; Prasad, S.K. IR regulation through preferential placement of h-BN nanosheets in a polymer network liquid crystal. *Mater. Horiz.* **2024**, *11*, 554–565. [CrossRef]
8. Iadlovsk, O.S.; Thapa, K.; Rajabi, M.; Mrukiewicz, M.; Shiyanovskii, S.V.; Lavrentovich, O.D. Electrically tunable total reflection of light by oblique helicoidal cholesteric. *MRS Bull.* **2024**, *49*, 835–850. [CrossRef]
9. Syrrakou, E.; Papaefthimiou, S.; Yianoulis, P. Eco-efficiency evaluation of a smart window prototype. *Sci. Total Environ.* **2006**, *359*, 267–282. [CrossRef]
10. Granqvist, C.G.; Wittwer, V. Progress in chromogenics: New results for electrochromic and thermochromic materials and devices. *Sol. Energy Mater. Sol. Cells* **2009**, *93*, 2032–2039. [CrossRef]
11. Wu, L.Y.; Zhao, Q.; Huang, H.; Lim, R.J. Sol-gel based photochromic coating for solar responsive smart window. *Surf. Coat. Technol.* **2017**, *320*, 601–607. [CrossRef]
12. Klaue, K.; Garmshausen, Y.; Hecht, S. Taking photochromism beyond visible: Direct one-photon nir photoswitches operating in the biological window. *Angew. Chem. Int. Ed.* **2018**, *57*, 1414–1417. [CrossRef] [PubMed]
13. Aburas, M.; Soebarto, V.; Williamson, T.; Liang, R.; Ebendorff-Heidepriem, H.; Wu, Y. Thermochromic smart window technologies for building application: A review. *Appl. Energy* **2019**, *255*, 113522. [CrossRef]
14. Wang, S.; Zhou, Y.; Jiang, T.; Yang, R.; Tan, G.; Long, Y. Thermochromic smart windows with highly regulated radiative cooling and solar transmission. *Nano Energy* **2021**, *89*, 106440. [CrossRef]
15. Yoo, S.H.; Park, M.K.; Park, J.S.; Kim, H.R. Enhanced adhesion and transmittance uniformity in laminated polymer-dispersed liquid crystal films. *J. Opt. Soc. Korea* **2014**, *18*, 753. [CrossRef]
16. Khandelwal, H.; Schenning, A.P.; Debije, M.G. Infrared regulating smart window based on organic materials. *Adv. Energy Mater.* **2017**, *7*, 1602209. [CrossRef]
17. Shaik, S.; Gorantla, K.; Mishra, S.; Kulkarni, K.S. Thermal and cost assessment of various polymer-dispersed liquid crystal film smart windows for energy efficient buildings. *Constr. Build. Mater.* **2020**, *263*, 120155. [CrossRef]

18. Li, C.C.; Tseng, H.Y.; Chen, C.W.; Wang, C.T.; Jau, H.C.; Wu, Y.C.; Hsu, W.H.; Lin, T.H. Versatile energy-saving smart glass based on tristable cholesteric liquid crystals. ACS Appl. *Energy Mater.* **2020**, *3*, 7601–7609.
19. Ji, S.M.; Oh, S.W.; Yoon, T.H. Smart window based on angular-selective absorption of solar radiation with guest–host liquid crystals. *Crystals* **2021**, *11*, 131. [CrossRef]
20. An, C.H.; Lee, J.H.; Han, C.H.; Oh, S.W. A smart window with angular-selective absorption for automotive. *Dyes Pigment.* **2023**, *212*, 111124. [CrossRef]
21. An, C.H.; Choi, T.H.; Oh, S.W. Energy-efficient liquid crystal smart window with a clear view. *Crystals* **2023**, *13*, 1464. [CrossRef]
22. Tukiainen, M. Sŏul, South Korea—Sunrise, Sunset, Dawn and Dusk Times for the Whole Year—Gaisma. Available online: https://www.gaisma.com/en/location/soul.html (accessed on 7 October 2024).
23. White, D.L.; Taylor, G.N. New absorptive mode reflective liquid-crystal display device. *J. Appl. Phys.* **1974**, *45*, 4718. [CrossRef]
24. Uchida, T.; Katagishi, T.; Onodera, M.; Shibata, Y. Reflective multicolor liquid-crystal display. *IEEE Trans. Electron Devices* **1986**, *33*, 1207–1211. [CrossRef]
25. Swinehart, D.F. The beer-lambert law. *J. Chem. Educ.* **1962**, *39*, 333. [CrossRef]
26. Scheffer, T.J. Optimized three-component dye mixtures for achromatic guest-host liquid-crystal displays. *J. Appl. Phys.* **1982**, *53*, 257. [CrossRef]
27. Yeh, P.; Gu, C. *Optics of Liquid Crystal Displays*, 2nd ed.; Wiley: New York, NY, USA, 2010; p. 297.
28. Oh, S.W.; Ji, S.M.; Han, C.H.; Yoon, T.H. A cholesteric liquid crystal smart window with a low operating voltage. *Dye. Pigment* **2021**, *197*, 109843. [CrossRef]

crystals

MDPI

Article

The Effects of Network Architecture on the Photomechanical Performance of Azo-Acrylate Liquid Crystal Elastomers

Anastasiia Svanidze [1], Sudarshan Kundu [1], Olena Iadlovska [1], Anil K. Thakur [1], Xiaoyu Zheng [2] and Peter Palffy-Muhoray [1,2,*]

1 Advanced Materials and Liquid Crystal Institute, Kent State University, Kent, OH 44242, USA; akthakur74@gmail.com (A.K.T.)

2 Department of Mathematical Sciences, Kent State University, Kent, OH 44242, USA

* Correspondence: mpalffy@kent.edu

Abstract: Azo-containing liquid crystal elastomers are photomechanical materials that can be actuated via illumination. The photomechanical response is a result of the photoisomerization of the azo moiety, which produces bulk stresses in the material. These stresses arise via two distinct and competing mechanisms: order parameter change induced stress and direct contractile stress. We describe thermomechanical and photomechanical experiments aimed at assessing the relative contributions of these. We show that the details of the attachment of azo dyes to the network can greatly influence the photoresponse. We discuss our results and summarize our findings.

Keywords: liquid crystal elastomers; thermal stress; photostress; azo; actuation; photomechanical

1. Introduction

Liquid crystal elastomers (LCEs) are a unique class of advanced materials that combine the properties of liquid crystals and rubber-like materials. They consist of a crosslinked polymer network polymerized with monomers possessing liquid crystalline phases. LCEs, first proposed by P.G. de Gennes [1] and realized by H. Finkelmann [2], are remarkable materials [3] due to their exceptional responsiveness to external stimuli, such as heating or illumination. This responsiveness originates from the sensitivity of liquid crystalline order to system parameters near a phase transition. LCEs hold significant potential for applications, as they can sustain and exert shear stresses as elastic solids. This makes them highly versatile for various applications requiring shape changes and stress response, such as soft robotics [4–7], artificial muscles [8,9], adaptive optics [10], and smart textiles [11–13], as well as biomedical applications [14–16]. The combination of mechanical flexibility and tunable properties makes LCEs well suited to next-generation responsive and adaptive technologies.

A great variety of LCEs has been synthesized, ranging from Finkelmann's early polyhydrosiloxane-based ones [2,17–22] to more recent acrylate-based [6,23–33], polyester-based materials [34,35], and others [36,37]. Actuation mechanisms used to change the liquid crystal order parameter include heating [6,8,15,23,24,26,27], photoactuation [3,4,7,17–22,25,28,30,31,38–50], the application of electric [51–55] and magnetic [56] fields, and chemical stimuli [57].

In LCEs, photoactuation can be achieved by incorporating azobenzene moieties into the polymer network. Upon illumination, the azobenzene undergoes photoisomerization, inducing internal stresses that can result in shape changes and enable the material to perform mechanical work. The photomechanical response can be attributed to two competing mechanisms: changes in the liquid crystal orientational order parameter and direct contractile stress exerted on the network.

Acrylate-based LCEs containing azobenzene moieties have attracted considerable interest in recent years [7,25,28,45–50]. In this work, we focus on amine–acrylate LCEs doped with azo dyes with different degrees of attachment to the network. We measured and compared the thermal and photoresponse of three types of samples: those in which the azobenzene moiety is covalently bonded into the network at both ends (2-azo), at one end (1-azo), and not bonded at all (free-azo). In this paper, we describe our sample materials, report experimental results, discuss our findings in light of the two competing mechanisms, and conclude by briefly summarizing our results.

2. Sample Composition and Structure

2.1. Materials

The chemical structure of the constituents of our samples is shown in Figure 1. The liquid crystal monomer 1,4-Bis[4-(6-acryloyloxyhexyloxy)benzoyloxy]-2-methylbenzene (RM82) was purchased from Jiangsu Hecheng Advanced Materials Co., Ltd. (Nanjing, China). The azo dyes 4,4′-Bis(6-acryloyloxyhexyloxy)azobenzene (2-azo) and acrylic acid (4-((4-hexyloxyphenyl) diazenyl)phenoxy)hexyl ester (1-azo) were purchased from SYNTHON Chemicals GmbH & Co., KG (Bitterfeld-Wolfen, Germany). The azo dye 1,2-Bis(4-(hexyloxy)phenyl)diazene (free-azo) was purchased, as a result of custom synthesis, from Henan Daken Chemical Co., Ltd. (Zhengzhou, China). The chain extender *n*-Butylamine and the photoinitiator phenylbis(2,4,6-trimethylbenzoyl)phosphine oxide (Irgacure 819) were purchased from Sigma-Aldrich, Merck KGaA (Darmstadt, Germany). All materials were used as received; no further chemical or structural characterization was carried out, and monomer purity was not taken into account in determining molar ratios. Polyimide SE-2170 (Nissan Chemical Corp., Tokyo, Japan) and its thinner 21 (Nissan Chemical Corp., Tokyo, Japan) for liquid crystal alignment were purchased from Brewer Science, Inc. (Rolla, MO, USA).

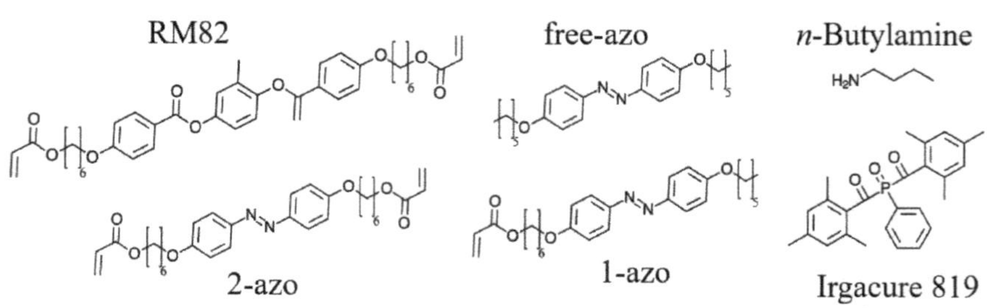

Figure 1. Chemical constituents of 2-azo, 1-azo, and free-azo samples.

Amine-acrylate-based LCEs were first proposed by T.J. White et al. [24]; a number of recipes are available for fabrication today [6,23–28]. Below, we briefly describe our fabrication process.

2.2. Network Synthesis and Sample Production

Plain glass substrates, 10 cm × 10 cm × 1 mm, were cleaned in an ultrasonic bath with detergent at 60 °C for 15 min, followed by the rinsing of substrates with deionized water and isopropanol. Following this, the substrates were dried in an oven for 15 min at 82 °C. The polyimide mixture SE-2170, in the proportion of 1:3 with its thinner, was spin-coated onto the clean substrates with the following sequence: 1 s at 500 rpm, 30 s at 1500 rpm, 1 s at 50 rpm. After spin-coating, the polyimide was soft-baked at 80 °C on a hotplate for 5 min and then hard-baked in an oven at 200 °C for 45 min. The glass substrates were then cooled to room temperature, and they were rubbed unidirectionally with velvet cloth on a rubbing block 10–15 times. They were subsequently cut into 4.5 cm × 4.5 cm squares and assembled, anti-parallel, into cells using Mylar film (DuPont Teijin Films, Chester, VA, USA) as spacers and Norland Optical Adhesive 68T (Norland Products Inc., Jamesburg, NJ, USA) as glue.

The liquid crystal monomer, azo dye, and photoinitiator were added to the amine and melted at 110 °C while stirring for 30 s. The mass of the photoinitator added was 2.5 wt.% of the combined mass of the other constituents. Relative concentrations of the components of samples are shown in Table 1.

Table 1. Compositions of sample types.

Sample Type	RM82 (Mole Fraction)	Azo Dye (Mole Fraction)	Acryl./Amine (Mole Ratio)	Irgacure 819 (wt.%)
no azo	1	0	1.1:1	2.5
free-azo 2%	0.98	0.02	1.1:1	2.5
1-azo 2%	0.98	0.02	1.1:1	2.5
2-azo 2%	0.98	0.02	1.1:1	2.5
free-azo 5%	0.95	0.05	1.1:1	2.5
1-azo 5%	0.95	0.05	1.1:1	2.5
2-azo 5%	0.95	0.05	1.1:1	2.5

Empty cells with planar surface alignment on the hotplate at 85 °C were filled with the mixture via the capillary effect. Filled cells were moved to an oven at 77 °C and left overnight for oligomerization.

The LCE cells were cured the following day using a mercury lamp (OmniCure s1500, Excelitas Technologies Corp., Pittsburgh, PA, USA). A filter with a cut-on wavelength of 400 nm (20CGA-400, Newport Corp., Irvine, CA, USA) was placed on each cell to reduce the photoisomerization of azobenzene during curing. Polymerization was carried out with light intensity of 75 mW/cm^2 at 70 °C for 20 min. The light intensity was measured at 405 nm with an Optical Power Meter (1830-R, Newport Corp., Irvine, CA, USA).

After curing, the cells were heated to 100 °C and opened using a thin blade. The LCE samples were removed from the substrate with tweezers.

The differences between samples were as follows: the 2-azo samples contained azo molecules with two covalently bondable double bonds, the 1-azo samples contained azo molecules with one covalently bondable double bond, and the free-azo samples contained azo molecules with no covalently bondable double bonds. All samples were produced using identical procedures; only the type (2-azo, 1-azo, and free-azo) and mole percentage (2% or 5%) of the azo compounds were changed.

A simple schematic of the resulting polymer network, shown in Figure 2, illustrates the key structural differences.

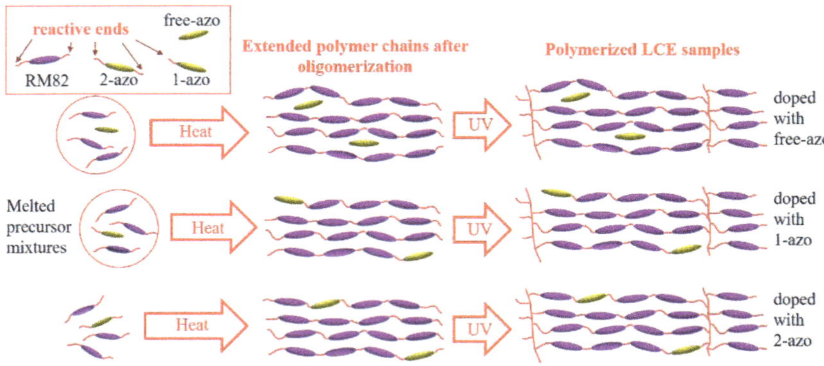

Figure 2. Networks of azo-doped acrylate-based LCEs.

3. Experimental Results

3.1. Elastic Moduli

Young's modulus of each samples was measured using a custom-built apparatus [29]. For these measurements, the LCE samples were cut into 4 mm × 20 mm strips with the director oriented parallel to the long edge. Samples marked with black dots were clamped at the top and bottom, and a tensile force was applied. A photograph of the sample was taken for each load, and the strain was determined using ImageJ software (Version 1.54f) [29]. A schematic of the device and experimental setup, together with typical data, is shown in Figure 3.

The LCE samples without any applied stress are not completely flat. Elongating a sample initially, therefore, requires very little stress, which then increases nonlinearly until the linear regime is reached. Measurements are shown in the linear regime. Data and Young's moduli for all samples are shown in Figure 4.

The elastic moduli measurements of Figure 4 are included to indicate that the elastic properties of our different samples are essentially identical, and differences in thermal and optomechanical properties do not originate from differences in elastic properties. It is also included to enable calculations of strains resulting from both initial stresses, as well as from thermal and photostresses measured in our experiments.

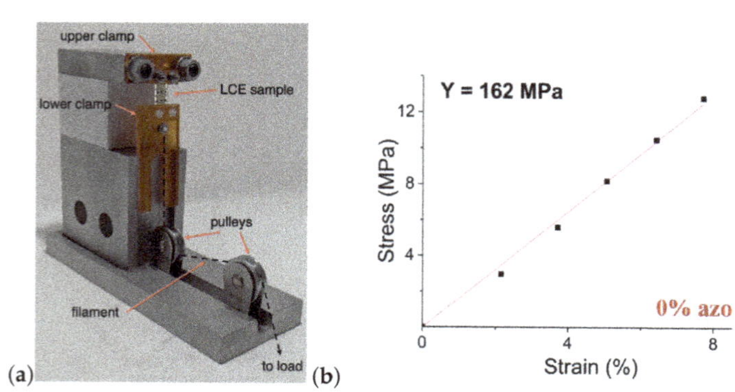

Figure 3. (**a**) Tensile force apparatus. (**b**) Young's modulus data for LCE sample with no azo dye.

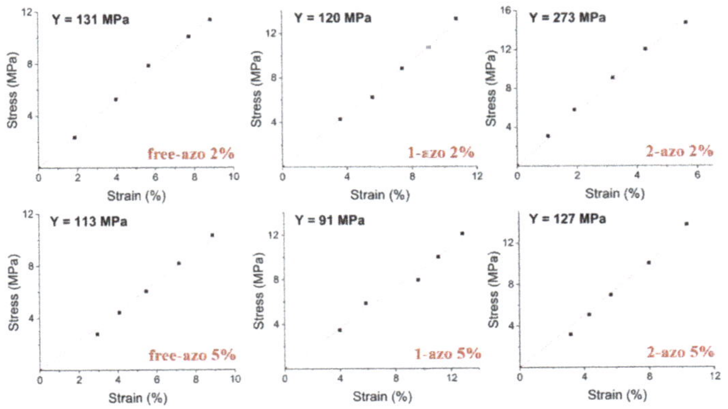

Figure 4. Young's modulus data for all samples. Top and bottom rows are for 2% and 5% free-azo, 1-azo, and 2-azo samples, respectively.

3.2. Thermal Stress Measurements

Our samples for thermal and photostress measurements were 30 mm long, 10 mm wide, and 27–32 μm thick, with nematic alignment along their long edge. They were held with the long edges vertical, and their two 10 mm edges were clamped between two rigid horizontal plexiglass struts, leaving a 10 mm × 10 mm sample area between holders. The bottom edge was held in a fixed position by a rigid support, while the top was connected to an Entran force sensor, whose height could be adjusted, with a nearly inextensible thin filament (Fireline jewelry thread). The sample and its holders were underwater in a thermostatted container. An image of the setup, used for both thermal and photostress measurements, is shown in Figure 5.

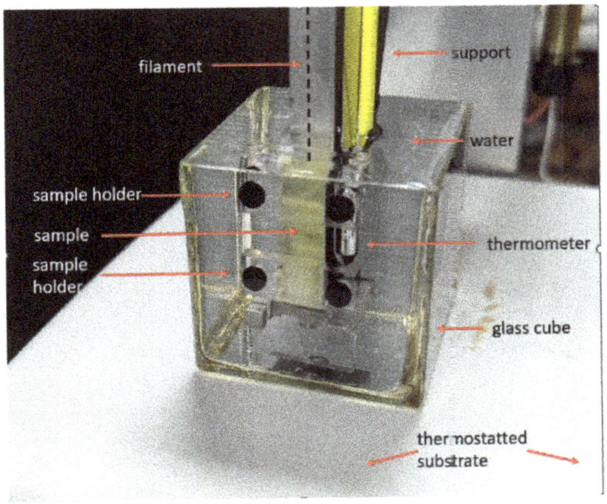

Figure 5. Thermal and photostress setup. The filament location is indicated by a dashed line. Thermal insulation was removed for visibility.

Thermal stress was determined via the following procedure. The initial stress on the sample was set by adjusting the height of the Entran sensor. The water temperature was increased in steps. At each temperature, the Entran voltage was recorded continuously for 1 min, and it was then averaged over time to provide the force and the stress exerted by the sample. The thermal stress was obtained by subtracting the initial stress from the measured stress. The heating rate between measurements was 0.25 °C/min.

Thermal stress data from azo-containing samples are presented below, together with photostress.

3.3. Transmission Spectra

Photoactuation in these samples is due to the photoisomerization of the azo moiety. When azo isomers, predominantly in the extended trans-configuration, absorb light at 365 nm, they undergo photoisomerization to the more compact cis-form, which has an absorption peak near 450 nm. Information about the dynamics of the population distribution can be inferred from transmission spectra.

We illuminated our samples with light from a 500 mW LED at 365 nm. The intensity of the illumination was 250 mW/cm^2 with duration 10 s. We then measured the transmittance spectra, whose minima provide information about the relative density of the trans- and cis-isomers, as a function of time. Transmission measurements were obtained using an HR4000CG-UV-NIR (Ocean Optics Inc., Orlando, FL, USA) spectrometer. Spectra were normalized with respect to the sample with no azo dye. The transmittance relaxation data are shown in Figure 6.

Material and optical parameters of the samples are given in Table 2. The absorption cross-section is $\sigma = 1/(\rho l_d)$, where ρ is the number density, and l_d is the decay length.

Although transmittance is clearly higher for the 2% than for the 5% azo containing samples, there is very little difference in the extent of isomerization in the free-azo, 1-azo, and 2-azo samples. We note here that the isomerization process is only weakly dependent on temperature, given the large excitation energy needed for azo trans–cis isomerization [30].

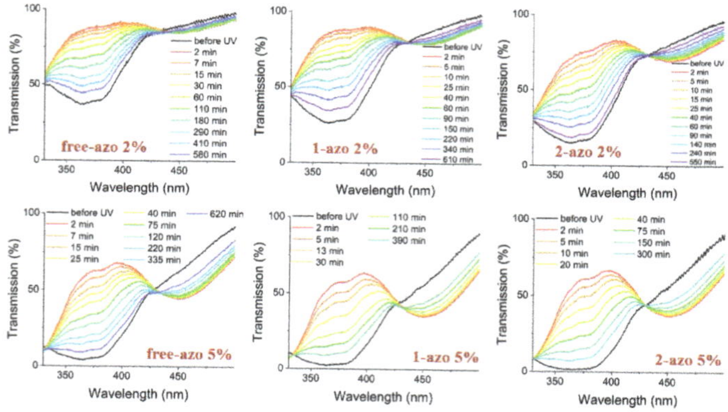

Figure 6. Transmission spectra of all samples. Top and bottom rows are for 2% and 5% of free-azo, 1-azo, and 2-azo samples, respectively.

Table 2. Sample material and optical parameters.

Sample	Thickness (µm)	Max Absorbance	Number Density of Dye (m^{-3})	Decay Length (µm)	Absorption Cross-Section (m^2)	Cis-Lifetime (s)
free-azo 2%	27	0.43	1.86×10^{25}	63	8.50×10^{-22}	1.2×10^4
1-azo 2%	27	0.57	1.94×10^{25}	47	1.10×10^{-21}	9.3×10^3
2-azo 2%	28	0.82	1.97×10^{25}	34	1.49×10^{-21}	4.4×10^3

Table 2. *Cont.*

Sample	Thickness (μm)	Max Absorbance	Number Density of Dye (m^{-3})	Decay Length (μm)	Absorption Cross-Section (m^2)	Cis-Lifetime (s)
free-azo 5%	31	1.36	4.70×10^{25}	23	9.20×10^{-22}	5.9×10^3
1-azo 5%	32	1.61	4.64×10^{25}	20	1.08×10^{-21}	6.5×10^3
2-azo 5%	27	1.87	4.84×10^{25}	14	1.48×10^{-21}	3.6×10^3

3.4. Photostress and Thermal Stress Measurements

Photostress measurements were carried out using the same setup as thermal stress measurements, shown in Figure 5. The sample was illuminated with UV light at 365 nm from a 500 mW LED (Prizmatix Ltd., Holon, Israel) at a distance of 6 cm from the sample. The light source was unpolarized, so polarization-dependent effects [31,43] need not be considered. The illuminated area of the sample was 10 mm × 10 mm. The illumination sequence consisted of the following: UV OFF for 10 s, then, repeatedly, UV ON for 200 ms, and UV OFF for 1 s. Since the cis-thermal relaxation time was very slow, as indicated in Table 2, after each measurement the sample was illuminated with light at 460 nm for 1 s to drive the cis–trans isomerization. The sample temperature as monitored via our thermometer was constant to within ±0.5 °C.

To illustrate that the photomechanical effect is a result of the presence of azo dyes, we show the thermal and photoresponse of samples without azo dye in Figure 7. There is no observable photostress in our thermostatted samples without azo dyes.

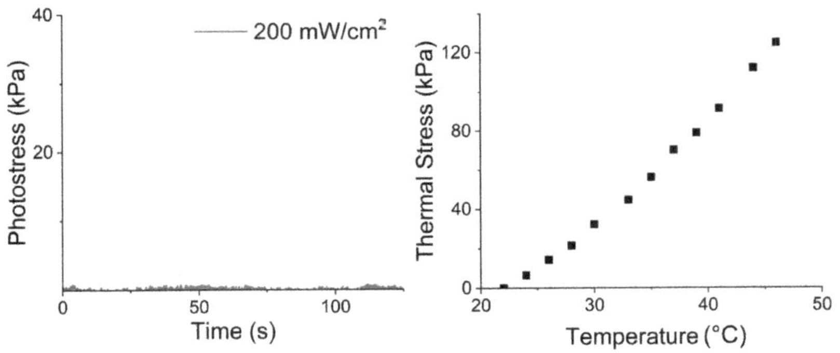

Figure 7. Photostress and thermal stress in acrylate LCE samples not containing azo dye. The initial stress is 50 kPa.

Photostress measurement results, together with those from thermal stress measurements, are shown in Figures 8 and 9. The results of measurements clearly indicate that the thermal stresses are essentially identical for free-azo, 1-azo, and 2-azo samples with the same initial stresses; however, photostresses of 2-azo samples differ dramatically from those of free-azo and 1-azo samples with the same initial stresses for both 2% and 5% azo concentrations. Our interpretations of these results are elucidated in Section 4.

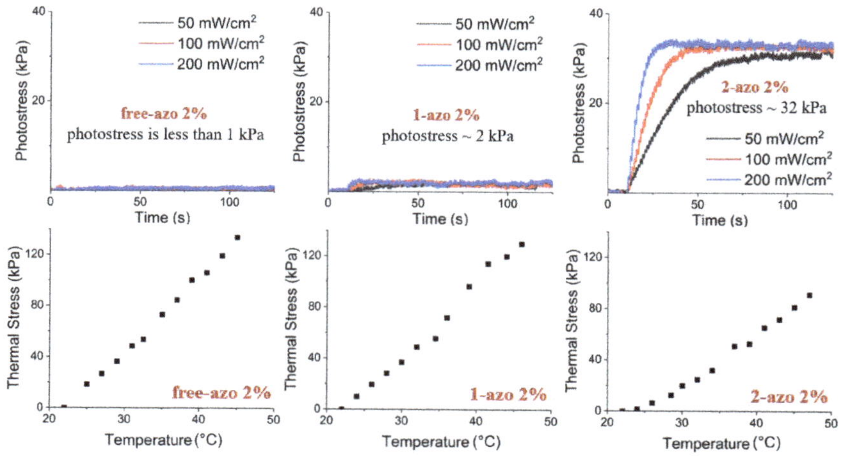

Figure 8. Photostress (**top**) and thermal stress (**bottom**) in acrylate LCE samples containing 2% free-azo, 1-azo, and 2-azo dye. The initial stress is 50 kPa.

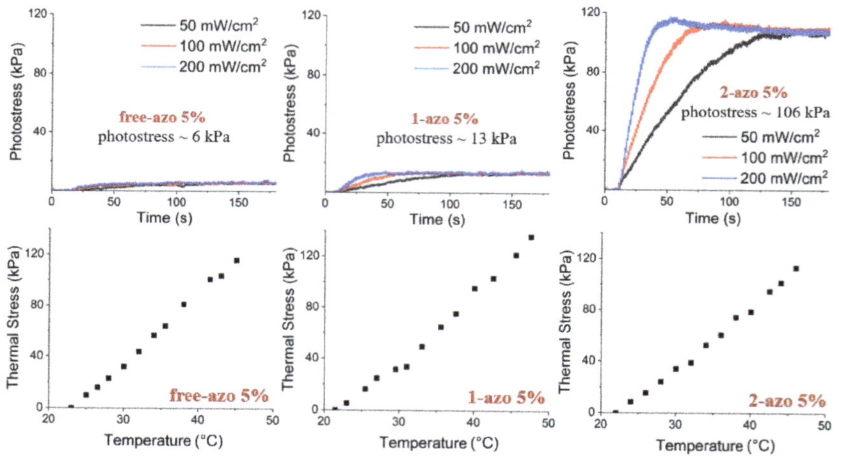

Figure 9. Photostress (**top**) and thermal stress (**bottom**) in acrylate LCE samples containing 5% free-azo, 1-azo, and 2-azo dye. The initial stress is 90 kPa.

4. Discussion

Heating changes the orientational order of nematic liquid crystals, which is necessarily coupled with mechanical strain [1,38]. This coupling persists in solid LC elastomers [3] and even in LC glasses [39]. Photoisomerization changes the shape of the azo-containing molecules. This gives rise to two competing mechanisms: 1. a change in orientational order due to the azo-containing molecules being less liquid crystal-like (the shape is less elongated, and the polarizability is less anisotropic) and 2. direct contractile stress due to the contraction of the two ends of the aligned azo-mesogens because of photoisomerization.

Perhaps these two mechanisms can be more clearly understood by considering the schematic of the network structure in Figure 2. The three rows of the right-hand column show the three network architectures studied. The elongated dark blue ellipsoids indicate the RM82 liquid crystal molecules. The extent of the nematic liquid crystal order, measured using the liquid crystal order parameter, is the extent to which these molecules are parallel.

In a pure liquid crystal, these molecules tend to be parallel both due to attractive London dispersion forces and due to shape-related steric interactions. These interactions are present in our samples, but they are modified by the topology of the network. Increasing the temperature reduces the degree of liquid crystalline order; the RM82 molecules become less parallel—the distance between neighbors along the chain is shortened, while the distance between neighbors along neighboring chains is increased. This is the essence of mechanism 1; this process was expected to be essentially the same in all three types of samples.

The effect of UV illumination is fundamentally different; it essentially causes the azo-containing molecules, represented by the elongated yellow ellipsoids, to change their shape and become shorter and wider due to photoisomerization. Although all azo-containing molecules are expected to undergo photoisomerization from the elongated trans- to the more compact cis-isomer shape, the effect on the network is dramatically different. In free-azo samples, shown in the top row, the shape change due to isomerization would have a slight effect on orientational order in the vicinity of the azo moiety, but only slight, since the concentration of the azo moiety is small. In one-azo samples, in the middle row, isomerization again will slightly reduce the orientational order, and due to the covalent bond at one end, it could give rise to a slight contractile force along the chain. However, in 2-azo samples, in addition to slightly reducing the orientational order, photoisomerization and the effective contraction of the molecule along its long axis will significantly shorten the chain, giving rise to a direct contractile force. This direct contractile force is the essence of mechanism 2.

The first of these, the change in the orientational order parameter, has often been referred to as the major mechanism for photoactuation in LCEs [17,31,38,40,42–44], and it has been seen as playing essentially the same role in both thermal and photoactuation [41,42,44]. The second, direct contractile stress, was already recognized by Finkelmann and Sanchez [20], who referred to the mechanism as a 'cooperative effect'. Earlier studies have probed the effects of different azo dye attachments [18,19], but the different structures of the dyes used made comparisons difficult. Later work on siloxane elastomers [20] with two different kinds of 2-azo and 1-azo dyes (also different from ours) indicated that higher photostress was exhibited by 2-azo samples. Although the justification is not clearly given in their paper, it is suggested that mechanism 2 provides 60%, and mechanism 1 40% of the stress.

Our three samples, with the 2-azo, 1-azo, and free-azo architectures, exhibit essentially identical thermal stress, as shown in the bottom rows of Figures 8 and 9. This implies that the differences in network architecture do not affect the thermally induced change in order parameter or the stress resulting from the order parameter change. The thermal stress is due to the change of order parameter caused by a change in temperature, and the stress is primarily the consequence of the coupling of the order parameter and the stress.

The three samples, with the 2-azo, 1-azo, and free-azo architectures, exhibit the same photoisomerization, as indicated by the spectroscopic results shown in Figure 6. If the photostress was primarily due to the change in order parameter via mechanism 1, then the photostress would be essentially the same for all samples since the photoisomerization is essentially the same. If the photostress was primarily due to a direct contractile stress due to mechanism 2, then the result for the 2-azo sample would be dramatically different from the response of the 1-azo and free-azo samples. Since the difference in the observed photostress is dramatically larger for the 2-azo samples than those for the 1-azo and free-azo samples, we conclude that the primary mechanism is the contractile stress due to photoisomerization in mechanism 2.

In light of the above, it is interesting to ask the following: what, then, is the role of liquid crystallinity in the photoresponse of 2-azo samples? Likely the role is to provide the alignment of the azo moiety when the sample is prepared. Acrylate LCEs, with or without azo compounds, are essentially aligned nematic samples, which would also align the predominantly trans-azo isomers. Our samples produce uniaxial stress because the stress-producing azo moiety is aligned, and the contraction of the unidirectionally aligned azo-nematogens, with their covalent bonds at two ends, efficiently transfer stress to the bulk network. It appears, then, that the main role of liquid crystallinity in the 2-azo samples is to provide the alignment of the stress-producing azo moieties for the production of unidirectional photostress.

5. Summary

Our experimental results indicate that photostress created in 2-azo acrylate LCEs is due to the direct unidirectional contraction of the network caused by the shape change of the azo moiety during photoisomerization. Stress associated with an order parameter change also exists, as observed in the 1-azo and free-azo samples, but it is about one order of magnitude smaller than photostress in the 2-azo materials. In the 2-azo samples, the traditional actuating mechanism via an order parameter change is, therefore, not significant; the primary role of liquid crystallinity is simply to align the photoresponsive azo-containing molecules during network formation.

Author Contributions: Conceptualization, P.P.-M. and X.Z.; investigation A.S., S.K., O.I. and A.K.T.; data curation A.S.; writing, P.P.-M., A.S. and X.Z.; funding acquisition P.P.-M. All authors have read and agreed to the published version of the manuscript.

Funding: We acknowledge support in part from the Office of Naval Research through the MURI on Photomechanical Material Systems (ONR N00014-18-1-2624).

Data Availability Statement: The data presented in this study are available on request from the corresponding author.

Acknowledgments: We are grateful to Dirk Broer and Danqing Liu for numerous enlightening discussions, as well as some sample materials. We are indebted to Bahman Taheri of AlphaMicron Inc., who kindly presented this work at ILCC2024 on our behalf.

Conflicts of Interest: The authors declare no conflicts of interest.

References

1. De Gennes, P.-G. Réflexions sur un type de polymères nématiques. *CR Acad. Sci. Ser. B* **1975**, *281*, 101–103.
2. Finkelmann, H.; Kock, H.-J.; Rehage, G. Investigations on Liquid-Crystalline Polysiloxanes. 3. Liquid-Crystalline Elastomers—A New Type of Liquid-Crystalline Material. *Macromol. Rapid Commun.* **1981**, *2*, 317–322. [CrossRef]
3. Warner, M.; Terentjev, E.M. Nematic Elastomers—A New State of Matter? *Prog. Polym. Sci.* **1996**, *21*, 853–891. [CrossRef]
4. Ahn, C.; Liang, X.; Cai, S. Bioinspired Design of Light-Powered Crawling, Squeezing, and Jumping Untethered Soft Robot. *Adv. Mater. Technol.* **2019**, *4*, 1900185. [CrossRef]
5. Wang, Z.; Guo, Y.; Cai, S.; Yang, J. Three-Dimensional Printing of Liquid Crystal Elastomers and Their Applications. *ACS Appl. Polym. Mater.* **2022**, *4*, 3153–3168. [CrossRef]
6. López-Valdeolivas, M.; Liu, D.; Broer, D.J.; Sánchez-Somolinos, C. 4D Printed Actuators with Soft-Robotic Functions. *Macromol. Rapid Commun.* **2017**, *39*, 1700710. [CrossRef]
7. Sartori, P.; Yadav, R.S.; del Barrio, J.; DeSimone, A.; Sánchez-Somolinos, C. Photochemically Induced Propulsion of a 4D Printed Liquid Crystal Elastomer Biomimetic Swimmer. *Adv. Sci.* **2024**, *11*, 2308561. [CrossRef]
8. Ma, J.; Wang, Y.; Sun, J.; Yang, Z. Liquid Crystal Elastomer Hollow Fibers as Artificial Muscles with Large and Rapid Actuation Enabled by Thermal-Pneumatic Enhanced Effect. *Adv. Funct. Mater.* **2024**, *34*, 2402403. [CrossRef]
9. Li, M.-H.; Keller, P. Artificial Muscles Based on Liquid Crystal Elastomers. *Phil. Trans. R. Soc. A* **2006**, *364*, 2763–2777. [CrossRef]

10. Ma, J.; Yang, Z. Chiral Liquid Crystal Elastomers Advance Light Modulation. *Light Sci. Appl.* **2024**, *13*, 205. [CrossRef]
11. Kim, S.-U.; Lee, Y.-J.; Liu, J.; Kim, D.S.; Wang, H.; Yang, S. Broadband and Pixelated Camouflage in Inflating Chiral Nematic Liquid Crystalline Elastomers. *Nat. Mater.* **2021**, *21*, 41–45. [CrossRef] [PubMed]
12. Roach, D.J.; Yuan, C.; Kuang, X.; Li, V.C.F.; Blake, P.; Romero, M.L.; Hammel, I.; Yu, K.; Qi, H.J. Long liquid crystal elastomer fibers with large reversible actuation strains for smart textiles and artificial muscles. *ACS Appl. Mater. Interfaces* **2019**, *11*, 19514–19521. [CrossRef] [PubMed]
13. Silva, P.E.S.; Lin, X.; Vaara, M.; Mohan, M.; Vapaavuori, J.; Terentjev, E.M. Active Textile Fabrics from Weaving Liquid Crystalline Elastomer Filaments. *Adv. Mater.* **2023**, *31*, 2210689. [CrossRef] [PubMed]
14. Prévôt, M.E.; Ustunel, S.; Hegmann, E. Liquid Crystal Elastomers—A Path to Biocompatible and Biodegradable 3D-LCE Scaffolds for Tissue Regeneration. *Materials* **2018**, *11*, 377. [CrossRef]
15. Wu, J.; Yao, S.; Zhang, H.; Man, W.; Bai, Z.; Zhang, F.; Wang, X.; Fang, D.; Zhang, Y. Liquid Crystal Elastomer Metamaterials with Giant Biaxial Thermal Shrinkage for Enhancing Skin Regeneration. *Adv. Mater.* **2021**, *33*, e2106175. [CrossRef]
16. Hussain, M.; Jull, E.I.; Mandle, R.J.; Raistrick, T.; Hine, P.J.; Gleeson, H.F. Liquid crystal elastomers for biological applications. *Nanomaterials* **2021**, *11*, 813. [CrossRef]
17. Cviklinski, J.; Tajbakhsh, A.R.; Terentjev, E.M. UV Isomerisation in Nematic Elastomers as a Route to Photo-Mechanical Transducer. *Eur. Phys. J. E* **2002**, *9*, 427–434. [CrossRef]
18. Hogan, P.M.; Tajbakhsh, A.R.; Terentjev, E.M. UV Manipulation of Order and Macroscopic Shape in Nematic Elastomers. *Phys. Rev. E* **2002**, *65*, 041720. [CrossRef]
19. Harvey, C.L.M.; Terentjev, E.M. Role of Polarization and Alignment in Photoactuation of Nematic Elastomers. *Eur. Phys. J. E* **2007**, *23*, 185–189. [CrossRef]
20. Sánchez-Ferrer, A.; Merekalov, A.; Finkelmann, H. Opto-Mechanical Effect in Photoactive Nematic Side-Chain Liquid-Crystalline Elastomers. *Macromol. Rapid Commun.* **2011**, *32*, 671–678. [CrossRef]
21. Dawson, N.J.; Kuzyk, M.G.; Neal, J.; Luchette, P.; Palffy-Muhoray, P. Modeling the Mechanisms of the Photomechanical Response of a Nematic Liquid Crystal Elastomer. *J. Opt. Soc. Am.* **2011**, *28*, 2134–2141. [CrossRef]
22. Campos, J.G.; Tobin, C.; Sandlass, S.; Park, M.; Wu, Y.; Gordon, M.; Read, J. Photoactivation of Millimeters Thick Liquid Crystal Elastomers with Broadband Visible Light Using Donor-Acceptor Stenhouse Adducts. *Adv. Mater.* **2024**, *36*, 24044932. [CrossRef]
23. Ware, T.H.; White, T.J. Programmed Liquid Crystal Elastomers with Tunable Actuation Strain. *Polym. Chem.* **2015**, *6*, 4835–4844. [CrossRef]
24. Ware, T.H.; McConney, M.E.; Wie, J.J.; Tondiglia, V.P.; White, T.J. Voxelated Liquid Crystal Elastomers. *Science* **2015**, *347*, 982–984. [CrossRef]
25. Ahn, S.; Ware, T.H.; Lee, K.M.; Tondiglia, V.P.; White, T.J. Photoinduced Topographical Feature Development in Blueprinted Azobenzene-Functionalized Liquid Crystalline Elastomers. *Adv. Funct. Mater.* **2016**, *26*, 5819–5826. [CrossRef]
26. Ambulo, C.P.; Burroughs, J.J.; Boothby, J.M.; Kim, H.; Shankar, M.R.; Ware, T.H. Four-Dimensional Printing of Liquid Crystal Elastomers. *ACS Appl. Mater. Interfaces* **2017**, *9*, 37332–37339. [CrossRef]
27. Kotikian, A.; Truby, R.L.; Boley, J.W.; White, T.J.; Lewis, J.A. 3D Printing of Liquid Crystal Elastomeric Actuators with Spatially Programed Nematic Order. *Adv. Mater.* **2018**, *30*, 1706164. [CrossRef]
28. Ceamanos, L.; Kahveci, Z.; López-Valdeolivas, M.; Liu, D.; Broer, D.J.; Sánchez-Somolinos, C. Four-Dimensional Printed Liquid Crystalline Elastomer Actuators with Fast Photoinduced Mechanical Response toward Light-Driven Robotic Functions. *ACS Appl. Mater. Interfaces* **2020**, *12*, 44195–44204. [CrossRef]
29. Oh, S.-W.; Guo, T.; Kuenstler, A.S.; Hayward, R.; Palffy-Muhoray, P.; Zheng, X. Measuring the Five Elastic Constants of a Nematic Liquid Crystal Elastomer. *Liq. Cryst.* **2020**, *48*, 511–520. [CrossRef]
30. Guo, T.; Svanidze, A.; Zheng, X.; Palffy-Muhoray, P. Regimes in the Response of Photomechanical Materials. *Appl. Sci.* **2022**, *12*, 7723. [CrossRef]
31. White, T. *Photomechanical Materials, Composites, and Systems: Wireless Transduction of Light into Work*; John Wiley & Sons, Ltd.: Hoboken, NJ, USA, 2017.
32. Berrow, S.R.; Raistrick, T.; Mandle, R.J.; Gleeson, H.F. Structure–Property Relationships in Auxetic Liquid Crystal Elastomers—The Effect of Spacer Length. *Polymers* **2024**, *16*, 1957. [CrossRef] [PubMed]
33. Yakacki, C.M.; Saed, M.; Nair, D.P.; Gong, T.; Reed, S.M.; Bowman, C.N. Tailorable and Programmable Liquid-Crystalline Elastomers Using a Two-Stage Thiol–Acrylate Reaction. *RSC Adv.* **2015**, *5*, 18997–19001. [CrossRef]
34. Yang, R.; Zhao, Y. Non-Uniform Optical Inscription of Actuation Domains in a Liquid Crystal Polymer of Uniaxial Orientation: An Approach to Complex and Programmable Shape Changes. *Angew. Chem. Int. Ed.* **2017**, *56*, 14202–14206. [CrossRef] [PubMed]
35. Beyer, P.; Braun, L.; Zentel, R. (Photo) Crosslinkable Smectic LC Main-Chain Polymers. *Macromol. Chem. Phys.* **2007**, *208*, 2439–2448. [CrossRef]

36. Lu, H.; Wang, M.; Chen, X.-M.; Lin, B.; Yang, H. Interpenetrating Liquid-Crystal Polyurethane/Polyacrylate Elastomer with Ultrastrong Mechanical Property. *J. Am. Chem. Soc.* **2019**, *141*, 14364–14369. [CrossRef] [PubMed]
37. Liu, X.-F.; Luo, X.; Liu, B.-W.; Zhong, H.-Y.; Guo, D.-M.; Yang, R.; Chen, L.; Wang, Y.-Z. Toughening Epoxy Resin Using a Liquid Crystalline Elastomer for Versatile Application. *ACS Appl. Polym. Mater.* **2019**, *1*, 2291–2301. [CrossRef]
38. Warner, M.; Terentjev, E.M. *Liquid Crystal Elastomers*; International Series of Monographs on Physics 120; Clarendon Press: Oxford, UK, 2007.
39. Warner, M.; Modes, C.D.; Corbett, D.P. Curvature in Nematic Elastica Responding to Light and Heat. *Proc. R. Soc. A Math.* **2010**, *466*, 2975–2989. [CrossRef]
40. Warner, M.; Terentjev, E. Thermal and Photo-Actuation in Nematic Elastomers. *Macromol. Symp.* **2003**, *200*, 81–92. [CrossRef]
41. Warner, M. Mechanical and Optical Bending of Nematic Elastomer Cantilevers. *Phys. Rev. E* **2012**, *86*, 022701. [CrossRef]
42. Warner, M.; Mahadevan, L. Photoinduced Deformations of Beams, Plates, and Films. *Phys. Rev. Lett.* **2004**, *92*, 134302. [CrossRef]
43. Corbett, D.; Warner, M. Nonlinear Photoresponse of Disordered Elastomers. *Phys. Rev. Lett.* **2006**, *96*, 237802. [CrossRef] [PubMed]
44. Corbett, D.; Warner, M. Changing Liquid Crystal Elastomer Ordering with Light—A Route to Opto-Mechanically Responsive Materials. *Liq. Cryst.* **2009**, *36*, 1263–1280. [CrossRef]
45. Tabrizi, M.; Clement, J.A.; Babaei, M.; Martinez, A.; Gao, J.; Ware, T.H.; Shankar, M.R. Three-Dimensional Blueprinting of Molecular Patterns in Liquid Crystalline Polymers. *Soft Matter* **2024**, *20*, 511–522. [CrossRef] [PubMed]
46. You, Y.; Golestani, Y.M.; Broer, D.J.; Yang, T.; Zhou, G.; Selinger, R.; Yuan, D.; Liu, D. Transforming Patterned Defects into Dynamic Poly-Regional Topographies in Liquid Crystal Oligomers. *Mater. Horiz.* **2024**, *11*, 3178–3186. [CrossRef] [PubMed]
47. Jiang, Z.-C.; Liu, Q.; Xiao, Y.-Y.; Zhao, Y. Liquid Crystal Elastomers for Actuation: A Perspective on Structure-Property-Function Relation. *Prog. Polym. Sci.* **2024**, *153*, 101829. [CrossRef]
48. Tholen, H.M.; Ambulo, C.P.; Lee, K.M.; Buskohl, P.R.; Harne, R.L. Optomechanical Computing in Liquid Crystal Elastomers. *Soft Matter* **2023**, *19*, 6978–6986. [CrossRef]
49. Van Raak, R.J.H.; Broer, D.J. Biomimetic Liquid Crystal Cilia and Flagella. *Polymers* **2022**, *14*, 1384. [CrossRef]
50. Lyu, P.; Astam, M.O.; Sánchez-Somolinos, C.; Liu, D. Robotic Pick-And-Place Operations in Multifunctional Liquid Crystal Elastomers. *Adv. Intell. Syst.* **2022**, *4*, 2200280. [CrossRef]
51. Fowler, H.E.; Rothemund, P.; Keplinger, C.; White, T.J. Liquid Crystal Elastomers with Enhanced Directional Actuation to Electric Fields. *Adv. Mater.* **2021**, *33*, 2103806. [CrossRef]
52. Urayama, K.; Honda, S.; Takigawa, T. Deformation Coupled to Director Rotation in Swollen Nematic Elastomers under Electric Fields. *Macromolecules* **2006**, *39*, 1943–1949. [CrossRef]
53. Choi, M.-Y.; Kim, K.; Kim, K.; Ahn, S.; Na, J.-H. Rapid and Spatially Programmed Electrostatic Actuation of Anisotropic Polymers. *Chem. Eng. J.* **2023**, *475*, 146237. [CrossRef]
54. Asgari, E.; Robichaud, A.; Cicek, P.-V.; Shih, A. Liquid Crystal Elastomers in Soft Micro Electromechanical Systems: A Review of Recent Developments. *J. Mater. Chem. C* **2024**, *12*, 15359–15381. [CrossRef]
55. Fowler, H.E.; Pearl, H.M.; White, T.J. Enhanced Electromechanical Output in Liquid Crystal Elastomers Prepared by Thiol-ene Photopolymerization. *Adv. Mater. Technol.* **2024**, *9*, 2301970. [CrossRef]
56. Winkler, M.; Kaiser, A.; Krause, S.; Finkelmann, H.; Schmidt, A.M. Liquid Crystal Elastomers with Magnetic Actuation. *Macromol. Symp.* **2010**, *291-292*, 186–192. [CrossRef]
57. Boothby, J.M.; Kim, H.; Ware, T.H. Shape Changes in Chemoresponsive Liquid Crystal Elastomers. *Sens. Actuators B Chem.* **2017**, *240*, 511–518. [CrossRef]

crystals

MDPI

Article

Photoinduced Interactions in Thin Films of Azo Dyes and Planar-Aligned Nematic Liquid Crystal

Aleksey Kudreyko [1,2,*], Vladimir Chigrinov [3,4] and Arina Perestoronina [1]

1 Department of General Physics, Ufa University of Science and Technology, Ufa 450076, Russia; arina.perest@yandex.ru
2 Department of Medical Physics and Informatics, Bashkir State Medical University, Ufa 450008, Russia
3 Department of Electronic & Computer Engineering, Hong Kong University of Science and Technology, Hong Kong 999077, China; eechigr@ust.hk
4 Department of Fundamental Physics and Nanotechnology, State University of Education, Moscow 105005, Russia
* Correspondence: akudreyko@uust.ru

Abstract: Properties of surface anchoring depend on the absorbed exposure energy and various potential interactions associated with liquid crystal and azo dye layers. In this study, we investigate a model of dispersion, steric and photoinduced interactions with the goal of providing a qualitative and quantitative description of orientationally ordered hard uniaxial liquid crystals and azo dye molecules. By using the Onsager theory, we estimated the effect of excluded volume. Typical repulsive potentials between liquid crystal and azo dye molecules are displayed graphically. The presence of statistical dispersion in molecular alignment of liquid crystals leads to potential wells in dipole–dipole interactions. Our mean field theory investigation of dipole–dipole interactions shows that the anchoring free energy is governed by the net interaction energy associated with the averaged dipole moments of liquid crystal and azo dye molecules, photoaligned surface dipole moments, and local charge densities. We also use the Fokker–Planck equation to show that rotational diffusion is described by the effective mean field potential, which includes photoinduced and van der Waals interactions. Our findings underscore the potential of mean field theory for intermolecular couplings in photoaligned surfaces, opening up new pathways of molecular design for a broad range of parameters.

Keywords: photoalignment; mean field theory; dipole–dipole interactions; steric interactions; Onsager theory; rotational diffusion

1. Introduction

Photoinduced alignment of liquid crystals (LCs) has been studied for decades [1,2]. Due to the light-induced rotational diffusion of dye molecules [3], the surface alignment of LC molecules, and the long-range ordering of LC mesophases, there exist many applications of photoalignment based on multi-domain structures [4–6]. Of all of the azo dye compounds encountered so far, sulphonic dye (SD1) has the 'best' properties for photoalignment [7]. Photoalignment is also a perfect effect for investigation of self-organized systems for experimental verification of theories [8,9]. The fundamental idea of the rotational diffusion model is the interaction potential between linearly polarized light and azo dye molecules, plus intermolecular van der Waals coupling, which acts as the thermodynamic average. Construction of a mathematical model that accounts for all possible interactions with statistical variations is rather complex, but we know that surface alignment of LCs and its stability depends on the shape of the surface anchoring potential [10,11]. Earlier measurements of photo-induced phase retardation have confirmed the model of rotational diffusion in the absence of LCs [3], but the interaction potential must also depend on long-range dipole–dipole coupling and short-range steric interactions. This defect in theory has been fairly noted in a number of studies, e.g., [12,13]. It now appears that the rotational diffusion mechanism should be extended to shape-changing molecules dissolved in biaxial LCs [14].

There exist a number of factors that affect alignment of LCs on azo dye layer. On one hand, chemical structure of LCs, e.g., homeotropic, planar alignment and pretilt angle of LCs depends on the length of the alkyl chain connecting the chalcone and polyimide main chains [15,16]. On the other hand, the chemical structure of azo dye determines the direction of its dipole moment. Combinations of various alignment methods also affect LC alignment [17,18].

The aim of our article is to describe energetically favored configurations of hard uniaxial LC and azo dye (AD) molecules by using interaction potentials and mean field theory. Currently known studies on rotational diffusion either do not consider dipole–dipole coupling and/or short-range steric interactions [3,19,20] or they rely on ad hoc linear vector differential equations describing surface alignment [21]. In order to make up for the missing estimations, we examine the impact of the mentioned couplings on the interaction potential.

Our approach is based on consideration of a LC cell which is sandwiched between photosensitive azo dye and photostable alignment layers. Typical values of intermolecular interaction energies range from 0.01–0.1 eV, and their nature is rather broad [22]. Referring to this range of energies, we apply interaction potentials for quantitative analysis of steric effects and dipole–dipole coupling between LC and AD molecules. Examination of steric effects reveals the role of AD molecular anisotropy and its contribution to the total interaction potential. By using the Onsager theory, we demonstrate that steric interactions between AD molecules are negligibly small. We also refer to the mean field Fokker–Planck equation to relate it to the rotational diffusion model and angular momentum operator. Our theoretical and computational results have a number of interesting consequences. For example, we explicitly show that the anchoring energy mainly depends on the mutual orientation of LC and AD dipole moments. We believe that our results are unbiased and consistent across a number of AD compounds with various geometrical and electrical properties.

2. Interaction Potential

Liquid crystal molecules are anchored to azo dye layer due to a number of interactions: light-induced rotational diffusion induces changes to dipole–dipole interactions between nematic LC and AD molecules and steric interactions. Formal representation of the corresponding potential has the following form:

$$U = U_{\text{PI}}(I, \Theta) + U_{\text{AD}}(\langle P_2 \rangle) + U_{\text{dd}}(\boldsymbol{r}) + U_{\text{GB}}(\mathbf{v}, \mathbf{u}, \mathbf{r}) \tag{1}$$

where $U_{\text{PI}}(I, \Theta)$ represents the photoinduced interaction potential between dye molecules and plane-polarized light. This term depends on the intensity of the irradiation light I and the angle Θ between the absorption oscillator and the polarization plane. At this point, it is necessary to specify that the orientations of long molecular axes and dipole moments do not coincide with each other. This complication can hardly be avoided; therefore, our model is described by numerous input parameters. Expression (1) can also be supplemented by photothermal effects, but it can be avoided by polymer stabilization of the azo dye film [23].

Rotational dynamics of light-sensitive molecules is described by the Smoluchowski equation or the Fokker–Planck equation [3,24]. Van der Waals interaction between uniaxial molecules with the thermodynamic mean molecular distribution is described by the

$U_{AD}(\langle P_2 \rangle)$ term, which also follows from the Fokker–Planck equation. Attractive electrostatic dipole–dipole (dd) interaction is accounted for by the $U_{dd}(\mathbf{r})$ term. Repulsive steric interactions between LC and dye molecules are described by the Gay–Berne potential (GB), represented by the U_{GB} ($\mathbf{v}$,$\mathbf{u}$,$\mathbf{r}$) term [22,25], where unit vectors $\mathbf{v}$ and $\mathbf{u}$ describe the orientations of LC and AD molecules, and r is the intermolecular vector. In the proceeding sections, we will consider the contribution of each term into the interaction potential U (Equation (1)).

2.1. Dispersion Interactions

Attractive long-range van der Waals and dipole–dipole interactions mediate anchoring effects. A well-known form of dipole–dipole interaction is the following:

$$U_{dd}(r) = \frac{1}{4\pi\varepsilon_0 r^3}[\boldsymbol{\mu}_{LC}\cdot\boldsymbol{\mu}_{AD} - 3(\boldsymbol{\mu}_{LC}\cdot\mathbf{R})(\boldsymbol{\mu}_{AD}\cdot\mathbf{R})], \quad (2)$$

where $\mathbf{R} = \boldsymbol{\rho} - \boldsymbol{\rho}'/r = \mathbf{r}/r$ represents a unit intermolecular vector between the centers of LC and AD molecules. Recall that LC molecules are aligned parallel to the azo dye molecules within the illuminated region [26]. This also means that dipole moments $\boldsymbol{\mu}_{AD}$ and surface LC molecules $\boldsymbol{\mu}_{LC}$ favor antiparallel alignment, tending to minimize the net dipole moment. It is convenient to define dipole moments by using unit vectors, i.e., $\boldsymbol{\mu}_{LC(AD)} = \mu_{LC(AD)}\mathbf{a}_{LC(AD)}$. Typical values of the pretilt angle are about 1° [27]. Then one can readily see that the expression for $U_{dd}(r)$ is simplified, and the interaction potential reads as follows:

$$U \approx \frac{1}{2}\alpha\tau I V_M \cos^2\Theta + a\langle P_2\rangle P_2(\Theta) - \frac{\mu_{LC}\mu_{AD}}{4\pi\varepsilon_0 r^3} + U_{GB}, \quad (3)$$

where α (1/m) is the absorption coefficient, τ is the relaxation time of AD, V_M is the volume of the AD molecule, and $a < 0$ (J) is a phenomenological constant. Equation (3) reveals that the lowest interaction energy between permanent molecular dipoles $U_{dd}(r)$ depends on dipole moments $\mu_{LC(AD)}$ and separation distance r [28,29]. As a result, dipole–dipole interactions for a perturbed molecular system must depend on the macroscopic polarization of LC and AD. This observation will be the subject of our further investigation.

A classic example that has been the subject of numerous studies is 4-cyano-4′-pentylbiphenyl (5CB) with the dipole moment $\mu_{LC} = 4$ D, which is applied from tail to cyano group [30]. The typical dipole moment of an AD molecule is $\mu_{AD} = 10$ D [31,32]. Then one can easily show that attractive dipole–dipole interactions $U_{dd}(r)$ range from $r \approx 4 - 10$ Å. In an idealized case, vector $\boldsymbol{\mu}_{AD}$ should be directed along the ITO-coated surface, i.e., belong to the xy-plane. One can assume that the orientation of LC dipole moments exhibits statistical dispersion along the alignment axis. This brings us to an important point of investigation of potential coupling between two groups of dipole moments.

Dipole–dipole interaction potential exhibits opposite signs depending on the mutual alignment of the two dipoles, described by polar $\theta_{LC(AD)}$ and azimuthal $\varphi_{LC(AD)}$ angles (see Figure 1a).

In order to perform calculations for estimation of the $U_{dd}(r)$ term, we represented expression (2) for $N = 10^3$ LC and AD molecules on a 9 to 1 basis. Azimuthal angle φ_{LC} was defined in terms of the Gaussian distribution as well as perfectly aligned LC molecules. Then the potential energy of the system was minimized under rigid-body constraints. As we expected, the calculated potential has repulsive and cohesive energies, and it is sensitive to the LC alignment quality, i.e., statistical distribution of φ_{LC}. Analysis of Figure 1b shows that antiparallel alignment of dipole moments minimizes the potential, but its behavior depends on how the LC azimuthal angle is defined. The existence of imperfect alignment of LC dipole moments leads to a potential well during the monotonous increase of angle φ_{LC}.

Consider the range $\pi \leq \varphi_{LC} \leq 3\pi/2$ when the anchoring effect is reduced through the potential well. The existence of statistical dispersion in φ_{LC} means that the total dipole moment of LC molecules possesses both parallel and perpendicular orientation components of polarization with the dominating longitudinal component.

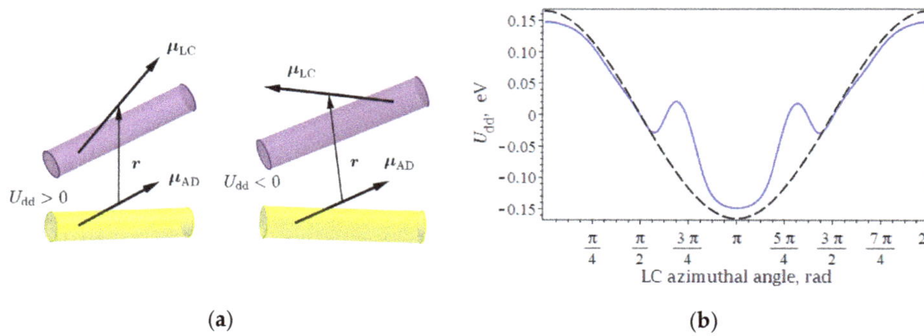

Figure 1. (**a**) Various configurations of dipole moments, providing repulsive and attractive interactions; (**b**) potential energy of pairwise dipole–dipole coupling versus φ_{LC} for normally distributed LC alignment (solid curve) and perfect alignment of LC molecules (dashed curve).

Continuous rotation of the dominating component μ_{LC} results in the rapid decay of the cohesive effect due to the presence of dipole moments, which are aligned in such a way as to increase the interaction energy. These molecules have a value of φ_{LC} greater than its mathematical expectation. With the increase of the mathematical expectation φ_{LC}, molecules from another 'tail' of the Gaussian curve have a net dipole moment that minimizes dipole–dipole coupling. As a result, we observe the potential well. When all LC dipole moments that are accounted for in the distribution form angles with AD dipole moments without oppositely directed components, the potential U_{dd} continues its monotonous increase to the amplitude value. Our calculations also show that the existence of the angular dispersion of LC dipole moments is also responsible for the decay of the amplitude value of U_{dd} due to the presence of multidirectional components of μ_{LC}.

To gain a better understanding of the net dipole–dipole interactions, consider a mean field approximation of expression (2), which reads as follows:

$$\langle U_{dd} \rangle = \frac{1}{2}\rho^2 \int \mathrm{H}(r-D) U_{dd}(r) f(\mathbf{a}_{LC}) f(\mathbf{a}_{AD}) r^2 dr\, \mathrm{d}^2\mathbf{R} \mathrm{d}\mathbf{a}_{AD} \mathrm{d}\mathbf{a}_{LC}, \quad (4)$$

where ρ is the number density, D is the distance between long molecular axes, $f(\mathbf{a}_{LC})$ and $f(\mathbf{a}_{AD})$ represent one-particle orientational distribution functions, and H is the Heaviside step function, which vanishes if AD and LC molecules penetrate each other; otherwise, it is equal to 1. In view of the molecular–statistical point of view, the integral over $\mathbf{R}$ in Equation (4) exhibits anomalous behavior in its volume integral due to the long-range nature of the potential. On one hand, the integral of the dipole–dipole potential over all orientations of the unit intermolecular vector $\mathbf{R}$ is equal to zero, as expected, in an isotropic matter:

$$\int U_{dd}(r) \mathrm{d}^2\mathbf{R} = 0.$$

On the other hand, the integral over $\mathbf{R}$ is logarithmically divergent at large intermolecular distances r. This means that the volume integral of the dipole–dipole potential is ill defined; i.e., the value of the integral depends on the system size, the shape of the sample, and boundary conditions [33]. However, the mean field approximation can be obtained by using Ewald's summation technique, which separates the short-range and long-range contributions of the dipole–dipole potential (2) and establishes the relationship that accounts for the electric field in the system [34]. As for the potential energy $U_{dd}(r)$, it must have a finite value, being electrically neutral.

Let us note that by combining Equations (2) and (4), we obtain average dipole moments: $\langle \mu_{LC} \rangle = \int \mathbf{\mu}_{LC} f(\mathbf{a}_{LC}) d\mathbf{a}_{LC}$ and $\langle \mu_{AD} \rangle = \int \mathbf{\mu}_{AD} f(\mathbf{a}_{AD}) d\mathbf{a}_{AD}$, where the mean quantities $\langle \mu_{LC} \rangle$ and $\langle \mu_{AD} \rangle$ are independent of the molecular size. Furthermore, $\frac{1}{2}\rho^2 \int U_{dd}(r) d\mathbf{a}_{AD} d\mathbf{a}_{LC} dV = -\frac{1}{2}\int \mathbf{P}(\mathbf{r})\mathbf{E}(\mathbf{r})d^3\mathbf{r}$, where $\mathbf{P}(\mathbf{r}) = \rho\langle \mathbf{\mu} \rangle$ is the macroscopic polarization. Then Expression (4), with the transformations of the mean field theory, represents the contribution to the anchoring energy, which depends on dipole moments and the shape of the sample:

$$\langle U_{dd} \rangle = \frac{4\pi}{3}\langle \mu_{AD} \rangle \langle \mu_{LC} \rangle - \frac{1}{2}\int \mathbf{P}(\mathbf{r})\mathbf{E}(\mathbf{r})d^3\mathbf{r}, \tag{5}$$

where the macroscopic electric field inside the pattern is related to cylindrically symmetric dipole–dipole tensor $\mathbf{T}$, i.e., $\mathbf{E}(\mathbf{r}) = -\int \mathbf{T}(\mathbf{\rho} - \mathbf{\rho}')\cdot\mathbf{P}(\mathbf{\rho}')d\mathbf{\rho}'$. This means that the weighted average in the PE term (5) contributes to the bulk electric field within the sample. The first term in Expression (5) contributes the anchoring free energy and is governed by the net interaction energy between LCs and the dipole moments of azo dye molecules and charge densities. The torques induced by interaction between the electric field of the exposure light and dipole moments $\mathbf{\mu}_{AD}$ provide two-dimensional control over its orientation.

2.2. Steric Interactions

With the particles of axial symmetry, the contribution of intermolecular steric interactions is well described by the GB potential and given as follows:

$$U_{GB} = \begin{cases} 4\varepsilon\left(A^{12} + A^6\right), \; r \le r_c, \\ 0, r > r_c, \end{cases} \tag{6}$$

where A is the range parameter, given by $A(\mathbf{v}, \mathbf{u}, \mathbf{R}) = \sigma_0/[r - \sigma(\mathbf{v}, \mathbf{u}, \mathbf{R}) + \sigma_0], \sigma(\mathbf{v}, \mathbf{u}, \mathbf{R}) = \sigma_0\left[1 - \chi/2(\mathbf{R}\cdot\mathbf{v} + \mathbf{r}\cdot\mathbf{u})^2/(1 + \chi(\mathbf{v}\cdot\mathbf{u})) + (\mathbf{R}\cdot\mathbf{v} - \mathbf{r}\cdot\mathbf{u})^2/(1 - \chi(\mathbf{v}\cdot\mathbf{u}))\right]^{-1/2}$ and $\varepsilon \equiv \varepsilon_0\left[1 - \chi^2(\mathbf{v}\cdot\mathbf{u})^2\right]^{-1/2}$ is the strength parameter; $\chi = \left(X_0^2 - 1\right)/\left(X_0^2 + 1\right)$ is the anisotropy parameter. Let these vectors be defined as follows: $\mathbf{v} = (\sin\theta\cos\varphi, \sin\theta\sin\varphi, \cos\theta)$ and $\mathbf{u} = (\cos\varphi, \sin\varphi, 0)$; the critical radius is defined as follows: $r_c = \left(2^{1/6} - 1\right)\sigma_0 + \sigma(\mathbf{v}, \mathbf{u}, \mathbf{R})$, where $\sigma_0 = \sqrt{2}a$. Here X_0 indicates the shape anisotropy parameter given by the ratio between the minor and major axes; i.e., $X_0 = b/a$. Let us remark that Model (6) suggests that the molecules are identical, which slightly contradicts the molecular lengths of azo dye and LC. Meanwhile, we will continue to use this model for estimation of azo dye–LC repulsive effects.

Surface arrangement of LCs and AD is driven by aromatic stacking interactions. Due to steric repulsions, adjacent molecular rods are expected to align in a random fashion. It is reasonable to consider the following types steric of interactions: AD-AD, LC-AD, and LC-LC and estimate the excluded volume effects between two molecules.

Let long molecular axes of azo dye and nematic LC molecules be represented as long cylinders; i.e., $d/l \ll 1$, where l is the length and d is the diameter. We note that steric coupling and dipole–dipole interaction are described by different principal axes defining long molecular axes $\mathbf{u}$, $\mathbf{v}$ and dipole moment $\mathbf{\mu}_{LC(AD)}$, which increases the complexity of our study. By adopting the Onsager theory [35,36], we can estimate the excluded volumes v_i, $i = 1,2,3$ for steric interactions:

$$\begin{aligned} v_1(\mathbf{u}_1, \mathbf{u}_2) &\approx 2d_{AD}l_{AD}^2|\mathbf{u}_1 \times \mathbf{u}_2|, \\ v_2(\boldsymbol{\nu}, \mathbf{u}) &\approx l_{LC}l_{AD}(d_{LC} + d_{AD})|\boldsymbol{\nu} \times \mathbf{u}|, \\ v_3(\boldsymbol{\nu}_1, \boldsymbol{\nu}_2) &\approx 2d_{LC}l_{LC}^2|\boldsymbol{\nu}_1 \times \boldsymbol{\nu}_2|. \end{aligned} \quad (7)$$

One can estimate the typical values of AD molecular length l_{AD} and diameter d_{AD} as 35 Å and 5 Å, respectively. Accordingly, LC molecular length l_{LC} and diameter d_{LC} are 20 Å and 5 Å. This enables us to calculate the excluded volumes for perpendicular alignment: $v_1 = 12.3$ nm^3, $v_2 = 7$ nm^3, and $v_3 = 4$ nm^3 (see Figure 2a). Here we can see that a relatively large length of an AD molecule leads to considerable values of the excluded volume. On the other hand, a low concentration ($\approx$ 1%) of AD molecules dissolved in N,N-Dimethylformamide will not have any effect on the excluded volume. In order to support our assumption, one can easily compare the total volume of a 1 nm-thick layer per 1 cm^2 (10^{-13} m^3) and the excluded volume of AD molecules in this bulk. Let the densities of N,N-dimethylformamide solution and SD1 be identical $\approx$ 940 kg/m^3; the molar mass of SD1 is known: 0.6865 kg/mol. Then this volume contains $\approx 8.24{\cdot}10^{11}$ molecules, which produce an excluded volume of $\approx 10^{-14}$ m^3. Therefore, steric interactions in the system of AD molecules do not have any significant effect. Meanwhile, the excluded volume effect in steric interactions between LC-AD molecules must play a dominant role because steric and dipole–dipole interactions between LC and AD molecules contribute to surface anchoring energy [22].

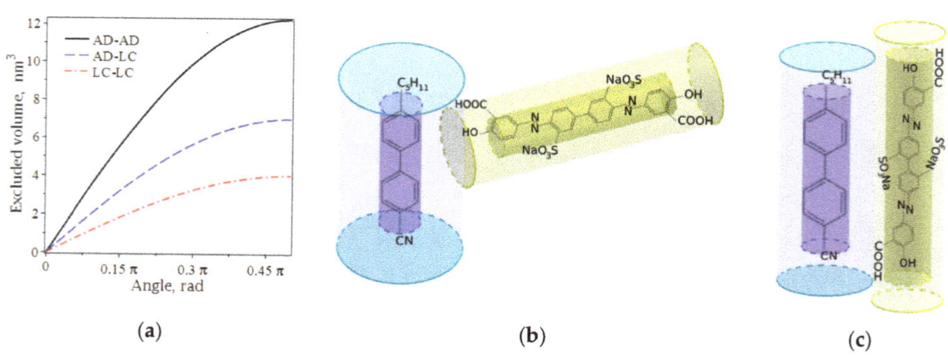

Figure 2. (**a**) Excluded volume versus principal axes, defining long molecular axes; illustration of the excluded volume effects for uniaxial molecules of LC and AD: (**b**) perpendicular and (**c**) parallel alignment. Chemical formulas in cylinder-shaped molecules represent azo dyes SD1m and SD1, respectively. Half-transparent cylinders denote the excluded volume.

In order to show the feasibility of various configurations between $\boldsymbol{\nu}$ and $\mathbf{u}$ vectors that provide the continuous dependence depicted in Figure 2a, consider NaO_3S groups in the benzene rings. Simulations in Vega ZZ show that the position of the NaO_3S group results in the changes in electron density, the spatial arrangement of atoms, and the direction of the net dipole moment with respect to the long molecular axis. An isomer of SD1 with the modified position of NaO_3S group (SD1m) possesses a dipole moment aligned perpendicular to its principal axis. Figure 2b shows the perpendicular arrangement of azo dye SD1m and nematic LC 5CB. The dipole moment of the well-known SD1 is applied along the molecular axis and results in the planar alignment of the LCs (see Figure 2c). Figure 2 also illustrates that perpendicular orientation of the rods results in a greater excluded volume (the area bounded by the shadowed surfaces) than parallel alignment (see Equation (7)).

By careful selection of the governing parameters in Equation (6), one can show that planar alignment ($\theta = 0$) with cluster formation is favored at the surface. A major explanation of this behavior is hydrogen bonding. In particular, surface coating with azo dye molecules (SD1) induces hydrogen bonds between $OH \cdots H$, $OH \cdots S$ and $NH \cdots O$ groups [37]. Hydrogen bonding between main chain structures of interacting molecules results in the formation of clusters. The characteristic time for H-bond formation/dissociation is about 10^{-12} s, which is much smaller than the thermodynamic relaxation time of SD1 (10^{-4} s) [3].

In order to obtain SD1-LC molecular system, two molecules were packed by using molecular visualization software PyMol 3.1.0. Then the two molecules were saved as a single file. Initial molecular system of many SD1-LC pairs was randomly generated by using Julia 1.11.2—platform runner for Packmol—software for initial molecular dynamics simulation [38]. Then we used the CHARMM force field, implemented in Vega ZZ 3.2.3.28, where hydrogen bonding is included implicitly in the electrostatic and van der Waals interactions. Our molecular simulation of the initial alignment shows that the azo dye molecules initially form clusters with the preferred alignment direction (see Figure 3). Further changes in molecular alignment are possible by photodissociation of hydrogen bonds. Then hydrogen bonds are formed again.

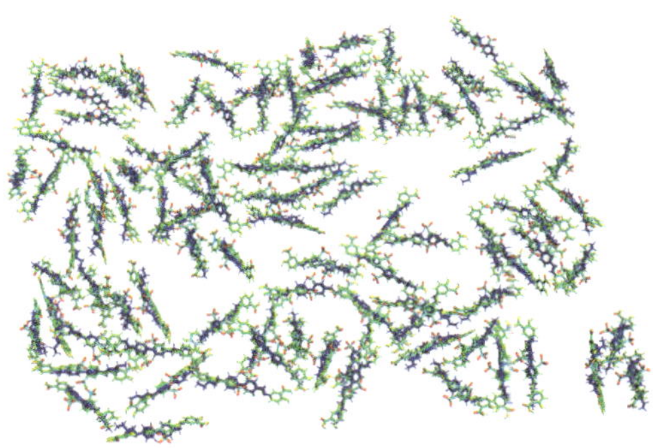

Figure 3. Simulation of 5CB (blue molecules) alignment along long molecular axis of SD1 (green molecules) on the substrate plane; surface size: 150 Å × 150 Å.

If the separation between molecular centers increases, it approaches to the critical radius r_c, when steric interactions become vanishingly small (see Figure 4). We also note that angle φ does not produce any contribution to steric interactions.

Investigation of the role of the anisotropy parameter X_0 on steric effects reveals an increase in U_{GB} potential for the perpendicular alignment of azo dye and LC molecules (see Figure 4b). Further assessment of the potential (6) versus separation distance r for various pretilt angles θ is depicted in Figure 4b. The separation distance that produces a repulsive potential of 0.01–0.1 eV represents a special interest because it corresponds to the typical intermolecular bonds that stabilize LC molecules on the surface. Note that we do not minimize the GB potential with respect to the intermolecular vector because there exist a number of other interactions. We only aim to make an assessment of the values of interaction energies.

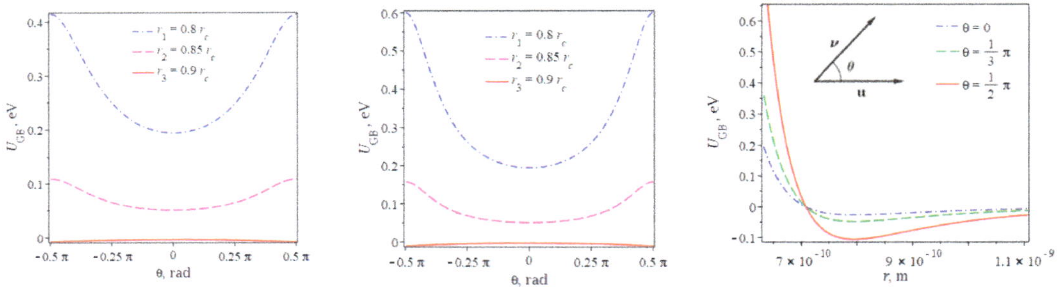

(**a**) (**b**) (**c**)

Figure 4. Repulsive effect of steric interaction for various distances between molecular centers of AD and LC for (**a**) $X_0 = 4$ and (**b**) $X_0 = 6$; (**c**) Gay−Berne potential versus separation distance between molecular centers of azo dye and LC for various angles θ.

Our assumption here is that the molecular anisotropy parameter X_0 is identical for AD and LC molecules, which might seem unrealistic, but it provides bold estimations of steric interactions for a number of compounds with various ratios between minor and major axes.

2.3. Photoinduced Potential

In order to reveal the nature of photoinduced interaction, let us refer to the Fokker–Planck equation. A key assumption taken in the diffusion models is that the angular distribution function $f(\mathbf{u}, t)$ of rod-like AD molecules must satisfy the rotational free energy:

$$\frac{\partial f}{\partial t} = -\mathbf{L}\cdot\mathbf{D}\left\{ f\mathbf{L}\frac{\delta F[f]}{\delta f} \right\}, \tag{8}$$

where $\mathbf{D}$ represents the rotational diffusion tensor and $\mathbf{L}$ is the angular momentum operator that can be expressed in terms of the Euler angles $(\theta_{AD}, \varphi_{AD})$ parametrization, i.e., differential operators have the form: $L_x = -i\left(\cos\theta_{AD}\cot\varphi_{AD}\frac{\partial}{\partial\theta_{AD}} - \sin\theta_{AD}\frac{\partial}{\partial\varphi_{AD}}\right)$, $L_y = -i\left(-\sin\theta_{AD}\cot\varphi_{AD}\frac{\partial}{\partial\theta_{AD}} + \cos\theta_{AD}\frac{\partial}{\partial\varphi_{AD}}\right)$, $L_z = -i\frac{\partial}{\partial\theta_{AD}}$. These simplified identities assume that the angular momentum operator is independent of the third Euler angle. Note that the momentum operator in (8) is expressed in terms of polar and azimuthal angles: $i\mathbf{L} = [\mathbf{r} \times \nabla]$, where $\mathbf{r} = \sin\theta_{AD}\cos\varphi_{AD}\mathbf{i} + \sin\theta_{AD}\sin\varphi_{AD}\mathbf{j} + \cos\theta_{AD}\mathbf{k}$. This substitution will lead to the replacement of spatial translations to rotations in the space of Euler angles. Since the effective free energy functional is the sum of the effective internal energy and the Boltzmann entropy, i.e., $F[f] = U[f] + \langle \ln f \rangle$, then the free energy term in Equation (8) can be written in a new form:

$$\frac{\partial f}{\partial t} = -\mathbf{L}\cdot\mathbf{D}\left\{ \mathbf{L}f + f\mathbf{L}\frac{\delta U}{\delta f} \right\}.$$

Therefore, rotational diffusion is governed by the effective mean field potential $\frac{\delta U}{\delta f}$. If out of plane reorientations are suppressed, then the mean field potential is equal to the photoinduced interaction and van der Waals interactions; i.e., $U_{PI}(I, \Theta) + a\langle P_2 \rangle P_2(\Theta)$ [22]. For rod-like SD1 molecules (see Figure 2) with cylindrical symmetry, only the angle Θ between the absorption oscillator and direction of the polarization plane matters. The selection rules of the absorption process dictate that the transition probability is proportional to $\cos^2\Theta$. As a result, rotational diffusion leads to an appearance of photoinduced dichroism and birefringence [3,39].

Let us estimate the term $U_{PI}(I, \theta)$. Typical values of dye molecular length l_{AD} and diameter d_{AD} constitute 35 Å and 5 Å, respectively. Then the molecular volume can be calculated as follows: $V_M = \pi d_{AD}^2 l_{AD}/4 \approx 0.7$ nm^3. Referring to the rotational diffusion model, let $\cos^2\Theta = 1$, $\alpha = 5{\cdot}10^7$ m^{-1} and $\tau = 5{\cdot}10^{-4}$ s; therefore $U_{PI}(I,\theta) \approx 0.1$ eV; i.e., photoinduced interaction potential is comparable to intermolecular interaction energies.

3. Discussion

We have found that dipole–dipole coupling potential between LC and AD molecules is sensitive to the angular distribution of dipole moments. The proposed model calculates the depth of the potential well in the net dipole–dipole interactions. Our theoretical calculations favor the presence of a potential well of $\cong 0.05$ eV to rotate a molecule of azo dye when LC and AD dipole moments are perpendicular to each other. This observation is consistent with the saturated dependence of the azimuthal anchoring energy of LC [26]. It is clear that the positive values of the U_{dd} term correspond to the repulsive effect; meanwhile, the cohesive effect between LC and AD molecules is also due to the impact of attractive van der Waals forces. This result can be treated as generalized, since we only considered dipole moments, not chemical structures.

Expressions for dipole–dipole and steric interactions (2) and (6) enable us to estimate these energies versus the direction of dipole moments and intermolecular distance. According to Onsager theory, these interactions must be supplemented by the effect of steric exclusion, described by Equation (7). However, the contribution of the dipole–dipole potential (2) to the free energy is ill defined and is not evaluated in a straightforward way.

The dipole moment vector μ_{AD} in AD molecule does not always coincide with its long molecular axis **u** (see Figure 1). In particular, the dihedral angle between ring planes is about 12°. This means that the total dipole moment has a number of projections, and LC molecules can be aligned to any energetically preferable position. In other words, central alignment as depicted in Figure 2 represents one of a number of possible configurations.

According to our qualitative estimations shown in Figures 1b and 4, the attractive contribution of van der Waals and dipole–dipole interactions is superior to the repulsive effect of LC-AD steric interactions, promoting planar alignment of LCs. Our further observation concerns Expressions (5) and (8). Light exposure induces rotational diffusion, which increases surface dipole moments $\langle\mu_{AD}\rangle$ and $\langle\mu_{LC}\rangle$ and anchoring energy. The anchoring energy achieves saturation when the dipole moments of AD and LC molecules have head-to-tail alignment. This is consistent with the saturated dependence of the azimuthal anchoring energy of nematic LCs [26].

4. Conclusions

In this article, we estimated the contribution of potential couplings during the rotational diffusion of azo dyes. Dipole–dipole interactions constitute the dominant factor of planar alignment of NLCs. Therefore, the investigation of the potential energy of pairwise dipole–dipole couplings versus the twist angle generalizes our results for other types of nematic LCs.

Development of mathematical formalism in dispersion interactions suggests that the interaction potential is equal to the photoinduced interaction and van der Waals interactions. The contribution of short-range steric interactions between azo dye molecules is negligibly small due to their low concentration. Investigation of steric and dipole–dipole interactions between AD and LC molecules can give indirect information to chemists and experimentalists regarding how to control attractive and repulsive forces.

Although our work is theoretical, it provides a currently unavailable framework for the analysis of potential interactions between photosensitive azo dye coatings and liquid crystals.

Author Contributions: Conceptualization: A.K. and V.C.; software and writing—original draft preparation: A.K. and A.P.; figures and editing: A.P. All authors have read and agreed to the published version of the manuscript.

Funding: The contribution of A. Kudreyko was supported by the Bashkir State Medical University Strategic Academic Leadership Program (PRIORITY-2030).

Data Availability Statement: All data are contained within the article.

Conflicts of Interest: The authors declare no conflict of interest.

References

1. Ichimura, K.; Suzuki, Y.; Seki, T.; Hosoki, A.; Aoki, K. Reversible change in alignment mode of nematic liquid crystals regulated photochemically by command surfaces modified with an azobenzene monolayer. *Langmuir* **1988**, *4*, 1214–1216. [CrossRef]
2. Ichimura, K. Photoalignment of Liquid-Crystal Systems. *Chem. Rev.* **2000**, *100*, 1847–1874. [CrossRef] [PubMed]
3. Chigrinov, V.; Pikin, S.; Verevochnikov, A.; Kozenkov, V.; Khazimullin, M.; Ho, J.; Huang, D.D.; Kwok, H.-S. Diffusion model of photoaligning in azo-dye layers. *Phys. Rev. E* **2004**, *69*, 061713. [CrossRef] [PubMed]
4. Nasrollahi, A.; Rella, A.K.; Kumar, V.; Kang, S.-W. Stepwise Progression of Dye-Induced In Situ Photoalignment and Subsequent Stabilization for Noncontact Alignment of Liquid Crystals. *ACS Appl. Mater. Interfaces* **2024**, *16*, 24052–24062. [CrossRef]
5. Quiroga, J.A.; Canga, I.; Alonso, J.; Crespo, D. Reversible Photoalignment of Liquid Crystals: A Path toward the Creation of Rewritable Lenses. *Sci. Rep.* **2020**, *10*, 1–9. [CrossRef]
6. Pan, J.-T.; Zhu, B.-H.; Ma, L.-L.; Chen, W.; Zhang, G.-Y.; Tang, J.; Liu, Y.; Wei, Y.; Zhang, C.; Zhu, Z.-H.; et al. Nonlinear geometric phase coded ferroelectric nematic fluids for nonlinear soft-matter photonics. *Nat. Commun.* **2024**, *15*, 8732. [CrossRef]
7. Folwill, Y.; Zeitouny, Z.; Lall, J.; Zappe, H. A practical guide to versatile photoalignment of azobenzenes. *Liq. Cryst.* **2021**, *48*, 862–872. [CrossRef]
8. Li, C.-Y.; Liu, S.-J.; Wu, H.-J.; Jiang, J.-Q.; Zhao, B.; Rosales-Guzmán, C.; Zhu, Z.-H.; Chen, P.; Lu, Y.-Q. Modal interface for structured light via liquid-crystal planar optics. *Phys. Rev. Appl.* **2024**, *21*, 034021. [CrossRef]
9. Nys, I.; Ropač, P.; Berteloot, B.; Ravnik, M.; Neyts, K. Highly dispersive liquid crystal diffraction gratings with continuously varying periodicity. *J. Mol. Liq.* **2023**, *383*, 122062. [CrossRef]
10. Ferrarini, A.; Moro, G.; Nordio, P.; Luckhurst, G. A shape model for molecular ordering in nematics. *Mol. Phys.* **1992**, *77*, 1–15. [CrossRef]
11. Dadivanyan, A.K.; Pashinina, Y.M.; Chausov, D.N.; Belyaev, V.V.; Solomatin, A.S. Mesogen Molecules Orientation on Crystal Surfaces. *Mol. Cryst. Liq. Cryst.* **2011**, *545*, 159–1383. [CrossRef]
12. Tavarone, R.; Charbonneau, P.; Stark, H. Kinetic Monte Carlo simulations for birefringence relaxation of photo-switchable molecules on a surface. *J. Chem. Phys.* **2016**, *144*, 104703. [CrossRef] [PubMed]
13. Geng, Y.; Yao, L. Effect of azimuthal anchoring energy on rewriting speed of optical rewritable e-paper. *Liq. Cryst.* **2021**, *48*, 915–921. [CrossRef]
14. de Souza, R.F.; Zannoni, C. Rotational diffusion of shape-switching particles dissolved in biaxial liquid crystals. *Liq. Cryst.* **2023**, *51*, 919–935. [CrossRef]
15. Jung, K.H.; Hyun, S.-Y.; Song, D.-M.; Shin, D.-M. The characteristics of polyimide photoalignment layer with chalcone derivatives produced by linear polarized UV light. *Opt. Mater.* **2003**, *21*, 663–666. [CrossRef]
16. Song, D.-M.; Jung, K.-H.; Moon, J.-H.; Shin, D.-M. Photochemistry of chalcone and the application of chalcone-derivatives in photo-alignment layer of liquid crystal display. *Opt. Mater.* **2003**, *21*, 667–671. [CrossRef]
17. Dogra, A.R.; Sharma, V.; Malik, P.; Kumar, P. In-situ homeotropic alignment of dye doped liquid crystal molecules on multilayered self assembled nanoparticles in confined cells for next generation display devices. *Opt. Mater.* **2024**, *154*, 115731. [CrossRef]
18. Ho, J.Y.L.; Chigrinov, V.G.; Kwok, H.S. Variable liquid crystal pretilt angles generated by photoalignment of a mixed polyimide alignment layer. *Appl. Phys. Lett.* **2007**, *90*, 2748345. [CrossRef]
19. Almasri, R.M.; Lim, S.S.; Lovell, N.H.; Ladouceur, F.; Al Abed, A. Alignment assessment of anisotropic liquid crystals through an automated image processing algorithm. *J. Mol. Liq.* **2024**, *408*, 125243. [CrossRef]
20. Ma, H.; Wereley, S.T.; Linnes, J.C.; Kinzer-Ursem, T.L. Computational fluid dynamics method for determining the rotational diffusion coefficient of cells. *Phys. Fluids* **2024**, *36*, 0193862. [CrossRef]
21. Berteloot, B.; Nys, I.; Liu, S.; Neyts, K. Two-Dimensional Liquid-Crystal Photoalignment by Multiple Illumination Steps. *ACS Appl. Opt. Mater.* **2023**, *2*, 1295–1302. [CrossRef]

22. Blinov, L.M. *Structure and Properties of Liquid Crystals*; Springer Science & Business Media: Berlin/Heidelberg, Germany, 2010; Volume 123.
23. Guo, Q.; Srivastava, A.; Chigrinov, V.; Kwok, H. Polymer and azo-dye composite: A photo-alignment layer for liquid crystals. *Liq. Cryst.* **2014**, *41*, 1465–1472. [CrossRef]
24. Kiselev, A.D.; Chigrinov, V.G.; Kwok, H.-S. Kinetics of photoinduced ordering in azo-dye films: Two-state and diffusion models. *Phys. Rev. E* **2009**, *80*, 011706. [CrossRef] [PubMed]
25. Gay, J.G.; Berne, B.J. Modification of the overlap potential to mimic a linear site–site potential. *J. Chem. Phys.* **1981**, *74*, 3316–3319. [CrossRef]
26. Chigrinov, V.G.; Kozenkov, V.M.; Kwok, H.-S. *Photoalignment of Liquid Crystalline Materials: Physics and Applications*; John Wiley & Sons: Hoboken, NJ, USA, 2008; Volume 17.
27. Ishihara, S.; Mizusaki, M. Alignment control technology of liquid crystal molecules. *J. Soc. Inf. Disp.* **2020**, *28*, 44–74. [CrossRef]
28. Osipov, M. Dipole-dipole interactions and the origin of ferroelectric ordering in polar nematics. *Liq. Cryst.* **2024**, 1–7. [CrossRef]
29. Dunmur, D.; Palffy-Muhoray, P. A mean field theory of dipole-dipole correlation in nematic liquid crystals. *Mol. Phys.* **1992**, *76*, 1015–1023. [CrossRef]
30. Thoms, E.; Yu, L.; Richert, R. From very low to high fields: The dielectric behavior of the liquid crystal 5CB. *J. Mol. Liq.* **2022**, *368*, 120664. [CrossRef]
31. Agwamba, E.C.; Udoikono, A.D.; Louis, H.; Udoh, E.U.; Benjamin, I.; Igbalagh, A.T.; Edet, H.O.; Ejiofor, E.U.; Ushaka, U.B. Synthesis, characterization, DFT studies, and molecular modeling of azo dye derivatives as potential candidate for trypanosomiasis treatment. *Chem. Phys. Impact* **2022**, *4*, 100076. [CrossRef]
32. Muravsky, A.; Murauski, A.; Kukhta, I.N. Photoinduced hole dipoles' mechanism of liquid crystal photoalignment. *Appl. Opt.* **2020**, *59*, 5102–5107. [CrossRef]
33. Groh, B.; Dietrich, S. Long-ranged orientational order in dipolar fluids. *Phys. Rev. Lett.* **1994**, *72*, 2422–2425. [CrossRef] [PubMed]
34. Wells, B.A.; Chaffee, A.L. Ewald summation for molecular simulations. *J. Chem. Theory Comput.* **2015**, *11*, 3684–3695. [CrossRef] [PubMed]
35. Onsager, L. The effects of shape on the interaction of colloidal particles. *Ann. N. Y. Acad. Sci.* **1949**, *51*, 627–659. [CrossRef]
36. Zannoni, C. *Liquid Crystals and Their Computer Simulations*; Cambridge University Press: Cambridge, UK, 2022.
37. O'Neill, M.; Kelly, S.M. Photoinduced surface alignment for liquid crystal displays. *J. Phys. D Appl. Phys.* **2000**, *33*, R67–R84. [CrossRef]
38. Martínez, L.; Andrade, R.; Birgin, E.G.; Martínez, J.M. Packmol: A package for building initial configurations for molecular dynamics simulations. *J. Comput. Chem.* **2009**, *30*, 2157–2164. [CrossRef]
39. Boychuk, A.; Shibaev, V.; Cigl, M.; Pomeisl, K.; Hamplová, V.; Pociecha, D.; Bubnov, A.; Bobrovsky, A. Photo-orientation Processes in Liquid Crystalline Polymethacrylates with Side Azobenzene Groups Having Lateral Methyl Substituents. *Macromolecules* **2021**, *54*, 10499–10509. [CrossRef]

crystals

MDPI

Article

Dissipation of Energy in a Compressible Nematic Microvolume Under Effect of a Temperature Gradient

Izabela Śliwa [1], Pavel V. Maslennikov [2] and Alex V. Zakharov [3,*]

1 Department of Operations Research and Mathematical Economics, Poznan University of Economics and Business, 61-875 Poznan, Poland; izabela.sliwa@ue.poznan.pl

2 Institute of Living Systems, Immanuel Kant Baltic Federal University, 236040 Kaliningrad, Russia; pashamaslennikov@mail.ru

3 Institute for Problems in Mechanical Engineering, Russian Academy of Science (IPME RAS), 199178 St. Petersburg, Russia

* Correspondence: alexandre.zakharov388@gmail.com

Abstract: We have carried out a numerical study of hydrodynamic processes in the hybrid-aligned channel of a compressible liquid crystal (HACLC) under the effect of a temperature gradient ∇T applied across a liquid crystal film. Calculations based on the classical Leslie–Ericksen theory showed that under the effect of ∇T, the HACLC sample settles down to a stationary flow regime with both horizontal u and vertical w components of velocity $\mathbf{v}$, and the direction and magnitude of $\mathbf{v}$ are strongly effected by the direction of ∇T. Calculations also showed that the relaxation of the stress tensor components $\sigma_{\mathrm{ij}}(z,t)$ $(\mathrm{i,j}=\mathrm{x,z})$ in the hybrid-aligned compressible nematic microvolume is determined by the direction and magnitude of the thermomechanical force.

Keywords: liquid crystals; microfluidics; hydrodynamics of anisotropic systems; thermomechanical effect

PACS: 61.30.Cz; 64.70.Md

1. Introduction

A liquid drop of macrosized volume placed between two horizontal solid surfaces, if heated from below, develops a buoyancy force due to the volume expansion of the fluid near the bottom. For sufficiently strong heating, when heat is transported only by heat conduction, the buoyancy force may overcome the viscous force and the liquid drop starts moving in the horizontal plane. One such example is the Rayleigh–Benard instability in a horizontal layer of anisotropic fluid heated from below. When the temperature gradient increases, at a critical value, and the static thermal diffusion can no longer transfer the thermal energy, then a macroscopic fluid flow occurs. This disturbance of mechanical balance occurs due to a local difference in buoyancy caused by the inhomogeneous thermal expansion of the fluid. As a result, dissipative roll patterns may occur in the horizontal layer. When the gradient is further increased, the flow develops into a nonperiodic flow. This dissipative system is characterized by complex interaction between the convective flow and structural deformations and the flowing material is not necessarily a simple molecular liquid but is often a complex fluid, such as the liquid crystal (LC) [1] or a living organism. Understanding the dissipation processes in LC systems under the influence of the temperature gradient is essential knowledge in material science and is of fundamental interest. While much is known in LC systems about dissipation processes in LC systems under the influence of both electric and magnetic fields, relatively little is known about the effect of these fields on the LC phases confined in a micro- or nanosized volume. The manipulation of tiny amounts of molecular liquids in channels ranging in size from tens to hundreds of micrometers has emerged as a separate field called microfluidics [2,3]. Microfluidics has become a paradigm in various fields from chemical synthesis and biological analysis to optics and information technology [4–6]. Central to the success of microfluidics is the development of innovative methods for the manipulation of LC systems in microchannels. Traditionally, an external electric field is used for controlling fluid motion [7–9]. On the other hand, microvolumes of LC materials are extremely sensitive to both the temperature gradient ∇T and boundary conditions, and, consequently, these factors, together with the anchoring forces, can influence the nature of hydrodynamic flow $\mathbf{v}$ in micro- or nanosized LC channels [10,11]. By the way, reducing the size of the channels to the nanoscale ensures a close connection between the nematic and biological fluids [4,6,12].

Recently, the problem of the motion of an ultra-thin (a few microliters) incompressible LC drop confined in the microsized volume between two horizontal and two lateral surfaces and subjected to uniform heating from below and above has been considered [10,11]. It has been shown that in the heat conduction regime the magnitude of the hydrodynamic flow $\mathbf{v}$ excited by ∇T in the hybrid-aligned incompressible nematic (HAIN) channel, is proportional $v \sim \frac{d}{\eta}\sigma_{zx}^{tm}$ to the tangential component of the thermomechanical stress tensor σ_{zx}^{tm}, where d and η are the thickness and viscosity of LC material [10,11], respectively. The direction of $\mathbf{v}$ is influenced both by the direction of the heat flux $\mathbf{q} = -T\frac{\delta\mathcal{R}}{\delta\nabla T}$ and the character of the preferred anchoring of the director $\hat{\mathbf{n}}$ to the restricted surfaces [10,11]. Here, $\mathcal{R}$ is the Rayleigh dissipation function of LC system.

In an attempt to make the next step towards the theoretical description of dissipation processes in confined compressible LCs under the influence of ∇T, we performed a numerical study of evolution not only of the stress tensor components of compressible fluid but also the redistribution of the velocity fields. To calculate it, one must include the equation of state and the equation for the velocity $\mathbf{v}$, as well as consider the coupled director–velocity equations in the framework of the well-established Leslie–Ericksen theory [13,14], as well as the thermoconductivity equation for the temperature field $T(\mathbf{r}, t)$ [15].

It should be noted that the horizontal layer of a resting LC fluid, heated from below, becomes unstable to convection due to the Rayleigh–Benard mechanism in the case when the driving force is not large enough to create convection. In this case, instability occurs at the value of $R = R_c \sim 1708$, regardless of the liquid under consideration [16]. Considering the size of an LC channel to be $\sim$ 10–20 μm, in our case $R \ll R_c$; in the future, we will focus primarily on the thermal conductivity mode as in HAIN and hybrid-aligned compressible nematic (HACN) channels.

The outline of this paper is as follows: The system of hydrodynamic equations describing both director motion and fluid flow of a compressible LC phase confined between two bounding surfaces, accounting for the heat conduction, caused by heating both from below or above, is given in Section 2. Numerical results for the relaxation regimes, caused by the vertical temperature gradient, describing orientational relaxation of the director, velocity, and stress tensor components, are given in Section 3. Conclusions are summarized in Section 4.

2. Formulation of the Balance Equations for Compressible Nematic Fluids

First of all, we would like to focus on describing the temperature T, density ρ, velocity $\mathbf{v}$, and director $\hat{\mathbf{n}}$ distributions that are produced in a two-dimensional (2D) microsized compressible nematic (CN) channel under the effect of a temperature gradient ∇T.

Consider the 2D HACN volume, which is bounded by two infinitely long boundaries at a distance of d on a scale of the order of tens micrometers and is initially at rest. Assuming that the temperature gradient ∇T changes only in the z direction, which coincides with the direction of the unit vector $\hat{\mathbf{k}}$ directed from the lower substrate to the upper one, we can assume that the components of the director $\hat{\mathbf{n}} = \sin\theta(z,t)\hat{\mathbf{i}} + \cos\theta(z,t)\hat{\mathbf{k}}$, as well as other physical quantities, depend only on the coordinate z and time t. Here θ denotes the polar angle, i.e., the angle between the direction of the director $\hat{\mathbf{n}}$ and the unit vector $\hat{\mathbf{k}}$, while the unit vector $\hat{\mathbf{i}}$ is directed parallel to the lower substrate, and $\hat{\mathbf{j}} = \hat{\mathbf{k}} \times \hat{\mathbf{i}}$, respectively. If the HACN volume, with the homeotropic aligning of the director field $\hat{\mathbf{n}}_{z=0} \parallel \hat{\mathbf{k}}$ on the lower, and planar $\hat{\mathbf{n}}_{z=d} \parallel \hat{\mathbf{i}}$, on the upper bounding surfaces is heated from below or above, due to the coupling between ∇T and $\nabla\hat{\mathbf{n}}$ in this LC volume, the hydrodynamic flow $\mathbf{v}(z,t) = v_x(z,t)\hat{\mathbf{i}} + v_z(z,t)\hat{\mathbf{k}} = u(z,t)\hat{\mathbf{i}} + w(z,t)\hat{\mathbf{k}}$ is exited.

The 2D hybrid alignment of the compressible nematic film confined between two solid surfaces suggests that

$$\theta(z)_{\mathrm{z=0}} = 0,\ \theta(z)_{\mathrm{z=d}} = \frac{\pi}{2}, \tag{1}$$

and its initial orientation is perturbed to be tilted with respect to the interface, with

$$\theta(z, t=0) = \frac{\pi}{2d}z, \tag{2}$$

and then allowed to relax to its equilibrium value $\theta_{\mathrm{eq}}(z)$. In turn, the velocity on these surfaces has satisfied the no-slip boundary condition

$$\begin{aligned} u(z)_{\mathrm{z=0}} = 0,\ u(z)_{\mathrm{z=d}} = 0, \\ w(z)_{\mathrm{z=0}} = 0,\ w(z)_{\mathrm{z=d}} = 0. \end{aligned} \tag{3}$$

Here, we consider two cases, firstly, when the upper surface is hotter than the lower one, and the boundary condition for the temperature field must satisfy

$$\begin{aligned} T_{\mathrm{z=d}} = T_{\mathrm{up}} = T_2,\ T_{\mathrm{z=0}} = T_{\mathrm{lw}} = T_1, \\ T_2 > T_1\ \text{(CASE I)} \end{aligned} \tag{4}$$

and secondly, when the lower surface is hotter than the upper one, and the boundary condition for the temperature field must satisfy

$$\begin{aligned} T_{\mathrm{z=d}} = T_{\mathrm{up}} = T_2,\ T_{\mathrm{z=0}} = T_{\mathrm{lw}} = T_1, \\ T_2 < T_1\ \text{(CASE II)}, \end{aligned} \tag{5}$$

respectively. Thus, solid surfaces are kept at different temperatures, and the HACN film is exposed to the temperature gradient ∇T directed parallel to the unit vector $\hat{\mathbf{k}}$.

Now, the reorientation of director field $\hat{\mathbf{n}}$ in the HACN film, confined between two solid surfaces, under the effect of the temperature gradient ∇T, when the relaxation regime is governed by viscous, elastic, and thermomechanical forces with accounting for hydrodynamic flow $\mathbf{v}$, can be obtained by solving a system consisting of mass, torque, momentum, and entropy balance equations.

If we assume that the liquid is compressible, then the mass balance equation will take the form

$$\rho_{,t} + \nabla \cdot (\rho\mathbf{v}) = 0, \tag{6}$$

where $\rho_{,t} = \partial\rho(z,t)/\partial t$. In turn, the torque balance equation gives us the equation for the reorientation of the director field, which in our case will take the form

$$\mathbf{T}_{\mathrm{el}} + \mathbf{T}_{\mathrm{vis}} + \mathbf{T}_{\mathrm{tm}} = 0, \tag{7}$$

where $\mathbf{T}_{\rm el} = \frac{\delta \mathcal{W}_{\rm el}}{\delta \hat{\mathbf{n}}} \times \hat{\mathbf{n}}$ is the elastic, $\mathbf{T}_{\rm vis} = \frac{\delta \mathcal{R}^{\rm vis}}{\delta \hat{\mathbf{n}}_{,t}} \times \hat{\mathbf{n}}$ is the viscous, and $\mathbf{T}_{\rm tm} = \frac{\delta \mathcal{R}^{\rm tm}}{\delta \hat{\mathbf{n}}_{,t}} \times \hat{\mathbf{n}}$ is the thermomechanical torques, respectively. Here, $\mathcal{W}_{\rm el}$ is the elastic energy, while $\mathcal{R}^{\rm vis}$ and $\mathcal{R}^{\rm tm}$ are the viscous and thermomechanical contributions to the total Rayleigh dissipation function. (For details, see Appendix A.)

The linear momentum equation for the velocity field $\mathbf{v}$ can be written as [10,11]

$$\rho \frac{d\mathbf{v}}{dt} = \nabla \cdot \sigma, \tag{8}$$

where the material derivative is $\frac{dv}{dt} = v_{,t} + ww_{,z}$, while the stress tensor (ST) $\sigma = \sigma^{\rm el} + \sigma^{\rm vis} + \sigma^{\rm tm} - P\mathcal{I}$ is the sum of elastic, viscous, thermomechanical parts and pressure and can be derived directly from the elastic contribution to energy and the Rayleigh dissipation function as $\sigma^{\rm el} = -\frac{\partial \mathcal{W}_{\rm el}}{\partial \nabla \hat{\mathbf{n}}} \cdot (\nabla \hat{\mathbf{n}})^{\rm T}$, $\sigma^{\rm vis} = \frac{\delta \mathcal{R}^{\rm vis}}{\delta \nabla \mathbf{v}}$, and $\sigma^{\rm tm} = \frac{\delta \mathcal{R}^{\rm tm}}{\delta \nabla \mathbf{v}}$, for the elastic, viscous, and thermomechanical contributions, respectively. Here, $w_{,z} = \partial w(z,t)/\partial z$ is the partial derivative of vertical component velocity with respect to z.

The viscous contribution to the total pressure P is given by $P = \sigma^{v} : \mathcal{I} = \sigma^{v}_{xx} + \sigma^{v}_{zz}$, while the elastic contribution to P is equal to $P^{el} = \sigma^{el} : \mathcal{I} = \sigma^{el}_{xx} + \sigma^{el}_{zz} = -2W_F$, where $W_F = \frac{1}{2}(K_1 n^2_{z,z} + K_3 n^2_{x,z})$, respectively. Here, $\mathcal{I}$ is the unit tensor, and equations for the tensor components σ^{v}_{xx} and σ^{v}_{zz} are given in Appendix A.

When the temperature gradient ∇T (~ 1.0 $[K/\mu m]$) is set up, we expect that the temperature field $T(z,t)$ satisfies the heat conduction equation [10,11]

$$C_{\rm p}(\rho T)_{,t} = -q_{z,z}, \tag{9}$$

where $C_{\rm p}$ is the heat capacity, $q_z = -T\frac{\delta \mathcal{R}}{\delta T_{,z}}$ is the heat flux in the HACN channel directed parallel to the unit vector $\hat{\mathbf{k}}$, while $q_{z,z} = \partial q_z/\partial z$, respectively. Taking into account that $\hat{\mathbf{n}} = \sin\theta(z,t)\hat{\mathbf{i}} + \cos\theta(z,t)\hat{\mathbf{k}}$ and $\mathbf{v}(z,t) = v_x(z,t)\hat{\mathbf{i}} + v_z(z,t)\hat{\mathbf{k}} = u(z,t)\hat{\mathbf{i}} + w(z,t)\hat{\mathbf{k}}$, the last equation can be rewritten in the form as

$$C_{\rm p}(\rho T)_{,t} = \lambda_{\perp}\Big[T_{,z}\Big(\lambda\cos^2\theta + \sin^2\theta\Big)\Big]_{,z} +$$
$$\xi\left[\theta_{,z}T\left(\theta_{,t}(\frac{1}{2} + \sin^2\theta) - u_{,z}\sin^2\theta(1 + \frac{1}{2}\sin^2\theta)\right)\right]_{,z} - \frac{\xi}{4}[Tw_{,z}\theta_{,z}\sin 2\theta]_{,z}, \tag{10}$$

where $\lambda = \lambda_{\parallel}/\lambda_{\perp}$ and $\xi \sim 10^{-12}$ J/mK is the thermomechanical constant [10,11]. Note that the density ρ, pressure P, and the temperature T of the liquid crystal system are connected by an equation of state, which in our case takes the form of a "Boussinesq approximation" [17]

$$\rho = \rho_0(1 - \alpha(T - T_1)), \tag{11}$$

where $\alpha = \frac{1}{\rho}\left(\frac{\partial \rho}{\partial T}\right)_P$ is the volume expansion coefficient and ρ_0 is the mass density of the LC phase.

In the case of the 2D compressible LC phase, the balance equations of dimensionless mass, torque, momentum, and entropy can be written inthe following forms. First of all, the condition of dimensionless compressibility (6) can be rewritten as

$$-\chi_{,\tau} - \chi_{,z}w - (\chi - \chi_1)w_{,z} + \frac{1}{\delta_1}w_{,z} = 0, \tag{12}$$

where $\chi(z,\tau) = T(z,\tau)/T_{\rm NI}$ is the dimensionless temperature, $T_{\rm NI}$ is the nematic–isotropic transition temperature, $\chi_1 = T_1/T_{\rm NI}$, $\tau = (K_{10}/\gamma_{10}d^2)t$ is the dimensionless time, $\delta_1 = T_{\rm NI}\alpha$ is the parameter of the LC system, while $\overline{z} = z/d$ and $\overline{w} = \frac{\gamma_{10}d^2}{K_{10}}w$ are the dimensionless distance away from the lower solid surface, and the vertical dimensionless component of velocity, respectively. Note that the overbars in the space variable z and the velocity component w have been eliminated. (For details, see Appendix A).

In turn, the dimensionless torque balance Equation (7) can be written as

$$\gamma_1(\chi)\theta_{,\tau} = \mathcal{A}(\theta)u_{,z} + (\mathcal{G}(\theta)\theta_{,z})_{,z} - \frac{1}{2}\mathcal{G}_{,\theta}(\theta)\theta^2_{,z} - \delta_2\chi_{,z}\theta_{,z}\left(\frac{1}{2} + \sin^2\theta\right), \tag{13}$$

where $\overline{\gamma}_1(\chi) = \gamma_1(\chi)/\gamma_{10}$ is the dimensionless rotational viscosity coefficient (RVC), $u_{,z} = \partial u(z,\tau)/\partial z$, $\mathcal{A}(\theta) = \overline{\mathcal{A}}(\theta)/\gamma_{10} = \frac{1}{2}(\gamma_1(\chi) - \gamma_2(\chi)\cos 2\theta)/\gamma_{10}$, $\mathcal{G}(\theta) = (K_1(\chi)\sin^2\theta + K_3(\chi)\cos^2\theta)/K_{10}$, $\mathcal{G}_{,\theta}(\theta)$ is the derivative of $\mathcal{G}(\theta)$ with respect to θ, and $\delta_2 = \xi T_{NI}/K_{10}$ is the parameter of the system, respectively. Here, $\xi \sim 10^{-12}\ J/mK$ is the thermomechanical constant [10,11], and the overbar in RVC $\overline{\gamma}_1(\chi)$ has been eliminated. (For details, see Appendix A).

In the case of compressible fluid, the dimensionless momentum balance Equation (8) reduces to

$$\delta_3[(1-\delta_1(\chi-\chi_1))u]_{,\tau} = \left[\overline{h}(\theta)u_{,z} - \mathcal{A}(\theta)\theta_{,\tau} - \delta_2\chi_{,z}\theta_{,z}\sin^2\theta\left(1+\frac{1}{2}\sin^2\theta\right) + \frac{\overline{\beta}_0}{6}w_{,z}\sin 2\theta\right]_{,z}, \tag{14}$$

and

$$\delta_3[(1-\delta_1(\chi-\chi_1))w]_{,\tau} = \left[\frac{\overline{\beta}_0}{6}\left(u_{,z}\sin 2\theta + \frac{2}{3}w_{,z}\right) - \frac{\delta_2}{4}\chi_{,z}\theta_{,z}\sin 2\theta - \frac{1}{2}\mathcal{G}(\theta)\theta^2_{,z}\right]_{,z}, \tag{15}$$

respectively. Here, $\delta_3 = \rho_0 K_{10}/\gamma^2_{10}$ is an additional parameter of the LC system. (For details, see Appendix A).

We also expect that the dimensionless temperature field $\chi(z,\tau)$ satisfies the heat conduction equation [10,11]

$$\delta_4[(1-\delta_5(\chi-\chi_1))\chi]_{,\tau} = \left[\chi_{,z}\left(\lambda\cos^2\theta + \sin^2\theta\right)\right]_{,z} + \delta_5\left[\chi\theta_{,z}\left(\theta_{,\tau}(\frac{1}{2}+\sin^2\theta) - u_{,z}\sin^2\theta(1+\frac{1}{2}\sin^2\theta)\right) - \frac{1}{4}w_{,z}\sin 2\theta\right]_{,z}, \tag{16}$$

where $\delta_4 = \rho_0 C_p K_{10}/(\gamma_{10}\lambda_\perp)$ and $\delta_5 = T_{\mathrm{NI}}K_{10}\xi/(\gamma_{10}\lambda_\perp d^2)$ are two extra parameters of the LC system. Note that the overbars in the space variable z in the last four Equations (13)–(16) have also been eliminated. (For details, see Appendix A).

Now, the reorientation of the director in the LC film confined between two solid surfaces, when the relaxation regime is governed by the viscous, elastic, and thermomechanical forces, and with accounting for the flow, can be obtained by solving the system of the nonlinear partial differential Equations (12)–(16), with the appropriate boundary conditions for the polar angle θ

$$\theta(z)_{z=0} = 0,\ \theta(z)_{z=1} = \frac{\pi}{2}, \tag{17}$$

with the no-slip boundary condition for both velocities

$$\begin{aligned} u(z)_{z=0} = 0,\ u(z)_{z=1} = 0, \\ w(z)_{z=0} = 0,\ w(z)_{z=1} = 0, \end{aligned} \tag{18}$$

and with the initial condition for the angle θ written in the form

$$\theta(z,\tau=0) = \frac{\pi}{2}z. \tag{19}$$

Below, two cases are considered: first, when the upper surface is hotter than the lower one, and the boundary condition for the temperature field must satisfy

$$\chi_{z=1} = \chi_{\rm up} = \chi_2,\ \chi_{z=0} = \chi_{\rm lw} = \chi_1,$$
$$\chi_2 > \chi_1 \text{ (CASE I)}, \tag{20}$$

and, second, when the lower surface is hotter than the upper one, and the boundary condition for the temperature field must satisfy

$$\chi_{z=1} = \chi_{\rm up} = \chi_2,\ \chi_{z=0} = \chi_{\rm lw} = \chi_1,$$
$$\chi_1 > \chi_2 \text{ (CASE II)}, \tag{21}$$

respectively.

For the case of $4-n-pentyl-4'-cyanobiphenyl$ $(5CB)$, at a temperature corresponding to the nematic phase, the values of the material constants are borrowed from the Ref. [10]. Thus, the mass density is $\sim 10^3\ kg/m^3$, $T_{\rm NI}(5CB) \sim 307$ K, and the highest values for elastic constants are $K_{10} \sim 13\ pN$, $K_{30} \sim 19\ pN$, and $\gamma_{10} \sim 0.077\ Pa\ s$, the specific heat $C_{\rm p} \sim 10^3\ J/kgK$, and the thermal conductivity coefficients $\lambda_{\parallel} \sim 0.24$ and $\lambda_{\perp} \sim 0.13\ W/mK$, respectively. The value of the thermomechanical constant ξ is equal to $\sim 10^{-12}\ J/m$ K, while the volume expansion coefficient α is equal to $\sim 10^{-3}\ K^{-1}$, respectively. Thus, the set of parameters values which are involved in Equations (12)–(16) are $\delta_1 \sim 0.3$, $\delta_2 \sim 24$, $\delta_3 \sim 2\times 10^{-6}$, $\delta_4 \sim 6\times 10^{-4}$, and $\delta_5 \sim 2\times 10^{-9}$, respectively. Using the fact that $\delta_3 \ll 1$, the momentum balance Equations (14) and (15) can be considerably simplified as velocities following adiabatically the motion of the director. Thus, the whole left-hand side of Equations (14) and (15) can be neglected, reducing it to

$$\sigma_{\rm zx} = \overline{h}(\theta)u_{,z} - \mathcal{A}(\theta)\theta_{,\tau} - \delta_2\chi_{,z}\theta_{,z}\sin^2\vartheta\left(1+\frac{1}{2}\sin^2\theta\right) + \frac{\overline{\beta}_0}{6}w_{,z}\sin 2\theta = C_1(\tau), \tag{22}$$

and

$$\sigma_{\rm zz} = \frac{\overline{\beta}_0}{6}\left(u_{,z}\sin 2\theta + \frac{2}{3}w_{,z}\right) - \frac{\delta_2}{4}\chi_{,z}\theta_{,z}\sin 2\theta - \frac{1}{2}\mathcal{G}(\theta)\theta^2_{,z} = C_2(\tau), \tag{23}$$

respectively, where the functions $C_1(\tau)$ and $C_2(\tau)$ does not depends on z and will be fixed by the boundary conditions. Equation (16) also can be considerably simplified, because both parameters $\delta_4, \delta_5 \ll 1$, and the whole left-hand side of Equation (16), as well as the second term, can be neglected so that Equation (16) takes the form

$$\left[\chi_{,z}\left(\lambda\cos^2\theta + \sin^2\theta\right)\right]_{,z} = 0. \tag{24}$$

The last equation has a solution

$$\chi(z,\tau) = \frac{\Delta\chi}{I}\int_0^z \left(\lambda\cos^2\theta + \sin^2\theta\right)^{-1} dz + \chi_i, \tag{25}$$

where $I = \int_0^1 \left(\lambda\cos^2\theta + \sin^2\theta\right)^{-1} dz$, $\Delta\chi = \chi_2 - \chi_1$, and $\chi_i = T_i/T_{\rm NI}$, respectively.

Now, let us consider the rest stress tensor components $\sigma_{\rm xz}$ and $\sigma_{\rm xx}$, respectively. Having obtained $\sigma_{\rm zx}$ (see, Equation (22)), and using the relation $\sigma_{\rm zx}(\tau) - \sigma_{\rm xz}(\tau) = \partial\mathcal{R}(\tau)/\partial\theta_{,\tau}$, one can calculate the dimensionless stress tensor component $\sigma_{\rm xz}$, which takes the form

$$\sigma_{\rm xz}(\tau) = \sigma_{\rm zx}(\tau) - \left(\mathcal{G}(\theta)\theta_{,z}\right)_{,z} + \frac{1}{2}\mathcal{G}_\theta(\theta)\theta^2_{,z}. \tag{26}$$

In the case of planar geometry, the dimensionless pressure $P(z,\tau) = \frac{d^2}{K_{10}}P(z,t)$ in the HACN channel is given by

$$P(z,\tau) = \sigma^{\rm vis}_{\rm zz}(z,\tau) = \sigma^{\rm vis}_{\rm xx}(z,\tau), \tag{27}$$

whereas the full form of $\sigma_{\rm xx}$ is given by

$$\sigma_{xx}(z,\tau) = \sigma_{xx}^{vis}(z,\tau) + \sigma_{xx}^{el}(z,\tau) = P(z,\tau), \tag{28}$$

which takes into account that $\sigma_{xx}^{el} = 0$. Here, $\sigma^{vis} = \frac{\delta \mathcal{R}^{vis}}{\delta \nabla \mathbf{v}}$ and $\sigma^{el} = -\frac{\partial \mathcal{W}_{el}}{\partial \nabla \hat{\mathbf{n}}} \cdot (\nabla \hat{\mathbf{n}})^{T}$ are the viscous and elastic contributions to the full stress tensor σ, respectively.

3. Orientational Relaxation of Velocity and Stress Tensor Components in the HACN Channel

The relaxation of director $\hat{\mathbf{n}}$ to its equilibrium orientation $\hat{\mathbf{n}}_{eq}$, which is described by the polar angle $\theta(z,\tau)$, from the initial condition $\theta(z,\tau=0) = \frac{\pi}{2}z$ to θ_{eq}, for both cases I and II, is shown in Figure 1a and 1b, respectively.

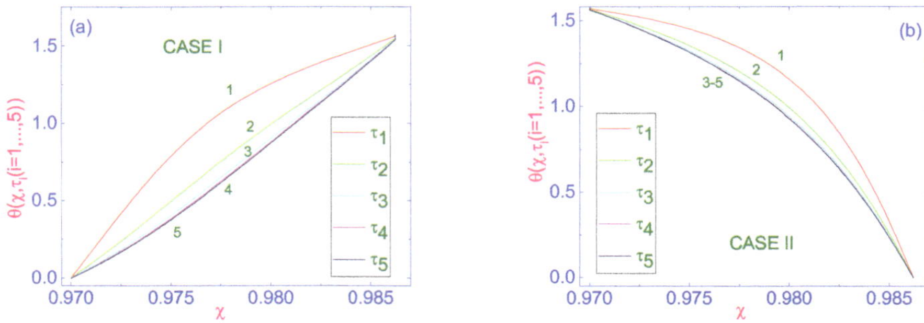

Figure 1. Temperature dependence of polar angle $\theta(\chi, \tau_i\ (i=1,\ldots,5))$, both for cases I (**a**) and II (**b**), and for the number of dimensionless times $\tau_i = \tau_0 i$, $(i=1,\ldots,5)$, whose values increase from curve (1) to curve (5). In case (**a**) (case I), the values of the temperature on both bounding surfaces are $\chi_{z=0} = \chi_1, \chi_{z=1} = \chi_1 + \Delta\chi$, while in case (**b**) (case II), these values are $\chi_{z=0} = \chi_1 + \Delta\chi, \chi_{z=1.0} = \chi_1$, respectively. Here, $\chi_1 = 0.97$, $\Delta\chi = 0.0162$ and $\tau_0 = 0.1$.

This was obtained by solving nonlinear partial differential Equations (13), (22), (23) and (25), together with the boundary conditions (17), (18), (20), and (21), as well as with the initial condition (19) by the numerical relaxation method [18]. In the calculations, the relaxation criterion $\epsilon = |(\theta_{(m+1)}(z,\tau) - \theta_{(m)}(z,\tau))/\theta_{(m+1)}(z,\tau)|$ was chosen to be equal to 10^{-4}, and the numerical procedure was then carried out until a prescribed accuracy was achieved. Here, m is the iteration number and τ_R is the relaxation time.

In case I, as shown in Figure 1a, the initial convex profile of $\theta(\chi,\tau_1)$ is converted into a set of concave profiles $\theta(\chi,\tau_i\ (i=2,\ldots,5))$, while in case II, as is shown in Figure 1b, the set of profiles $\theta(\chi,\tau_i\ (i=1,\ldots,5))$ is described by convex curves. Later, this difference in the behavior of $\theta(\chi,\tau_i\ (i=1,\ldots,5))$ profiles will be reflected in the different behavior of the velocity profiles, because we will be dealing with a different behavior of $\nabla\hat{\mathbf{n}}$.

The horizontal component of velocity $v_x(z,\tau_i\ (i=1,\ldots,5)) \equiv u(z,\tau_i\ (i=1,\ldots,5))$ as a function of the dimensionless distance z counted from the lower cooler $\chi_{z=0} = \chi_1$ to the upper warmer $\chi_{z=1} = \chi_1 + \Delta\chi$ (case I) bounding surface, for the number of dimensionless times $\tau_i = \tau_0 i$, $(i=1,\ldots,5)$, the values of which increase from the curve (1) to the curve (5), is shown in Figure 2a, while the same dependencies for case II are shown in Figure 2b, respectively.

In case I, when the LC sample is heated from above with the dimensionless temperature difference $\Delta\chi = 0.0162$ ($\sim$ 5 K), the profiles of the horizontal component $u(z,\tau_i\ (i=1,\ldots,5))$ of velocity are characterized by monotonic decreasing of $u(z,\tau_i\ (i=1,\ldots,5))$ with changing τ, before getting the equilibrium distribution $u_{eq}(z)$ across the LC channel. This distribution is characterized by the local maximum near the cooler lower boundary $u(z=0.27,\tau_5) = u_{eq}(z=0.27) = 1.94$ (35.5 μm/s), and the horizontal component of velocity is directed in the negative direction. In case II, the evolution of velocity profiles is more complicated. In this case, as shown in Figure 2b, the initially concave profile of $u(z, \tau_1)$, with the horizontal velocity component directed in the negative direction is transformed into a set of convex profiles $u(z, \tau_i\ (i = 2, \ldots, 5))$, with the horizontal component of the velocity directed in the positive direction. This distribution is characterized by the local maximum near the center of the LC sample $u(z = 0.52, \tau_5) = u_{\text{eq}}(z = 0.52) = 0.36$ (6.6 μm/s), and the horizontal component of the velocity is directed in the positive direction. A simple comparison of the maximum values of $u_{\text{eq}}(z = 0.27) = 1.94$ (35.5 μm/s), in case I, and $u_{\text{eq}}(z = 0.52) = 0.36$ (6.6 μm/s), in case II, shows that in the first case, this value is almost 5 times greater than in the second case.

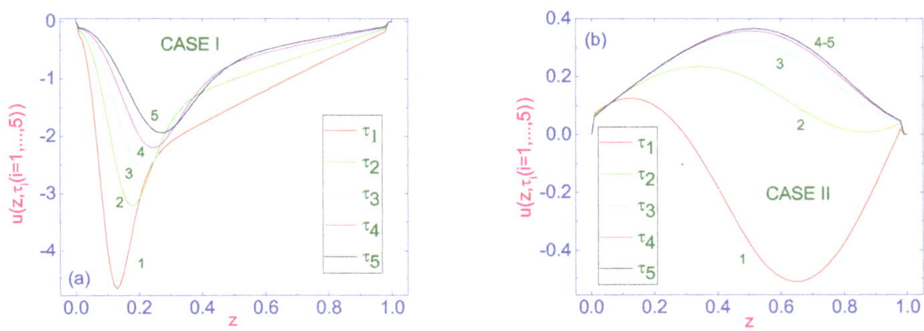

Figure 2. (**a**) The horizontal component of velocity $u(z, \tau_i)$ $(i = 1, \ldots, 5)$ as a function of the dimensionless distance z counted from the lower cooler $\chi_{z=0} = \chi_1$ to the upper warmer $\chi_{z=1} = \chi_1 + \Delta\chi$ (case I) bounding surface, for the number of dimensionless times $\tau_i = \tau_0 i$, $(i = 1, \ldots, 5)$, whose values increase from curve (1) to curve (5). Here $\chi_1 = 0.97$, $\Delta\chi = 0.0162$ and $\tau_0 = 0.1$. (**b**) Same as in (**a**), but for case II.

In turn, the vertical component of velocity $v_z(z, \tau_i\ (i = 1, \ldots, 5)) \equiv w(z, \tau_i\ (i = 1, \ldots, 5))$ as a function of the dimensionless distance z counted from the lower cooler $\chi_{z=0} = \chi_1$ to the upper warmer $\chi_{z=1} = \chi_1 + \Delta\chi$ (case I) bounding surface, for the number of dimensionless times $\tau_i = \tau_0 i$, $(i = 1, \ldots, 5)$, the values of which increase from the curve (1) to the curve (5), are shown in Figure 3a, while the same dependencies, for case II are shown in Figure 3b, respectively.

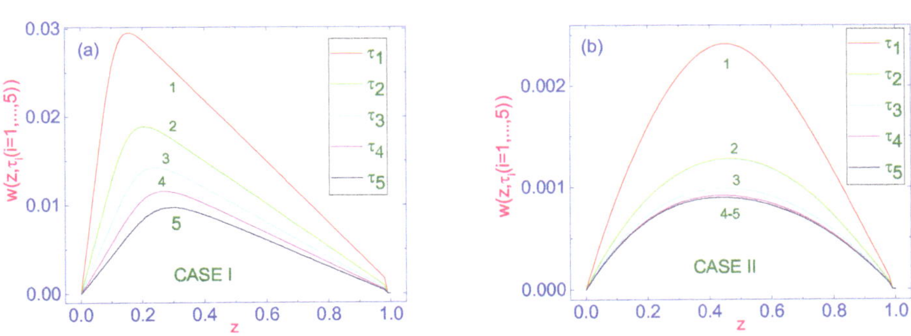

Figure 3. The vertical component of velocity $w(z, \tau_i)$ $(i = 1, \ldots, 5)$ as a function of the dimensionless distance z counted from the lower cooler $\chi_{z=0} = \chi_1$ to the upper warmer $\chi_{z=1.0} = \chi_1 + \Delta\chi$ (case I) bounding surface, for the number of dimensionless times $\tau_i = \tau_0 i$, $(i = 1, \ldots, 5)$, whose values increase from curve (1) to curve (5). Here $\chi_1 = 0.97$, $\Delta\chi = 0.0162$ and $\tau_0 = 0.1$. (**b**) Same as in (**a**), but for case II.

In case I, when the LC sample is heated from above with the dimensionless temperature difference $\Delta\chi = 0.0162$ (∼5K), the profiles of the vertical component

$w(z, \tau_i\ (i = 1, \ldots, 5))$ of velocity are characterized by monotonic decreasing of $w(z, \tau_i\ (i = 1, \ldots, 5))$ with changing τ, before getting the equilibrium distribution $w_{eq}(z)$ across the LC channel. This distribution is characterized by the local maximum near the cooler lower boundary $w(z = 0.3, \tau_5) = w_{eq}(z = 0.3) = 0.01$ (0.2 μm/s), and the vertical component of velocity is directed in the positive direction. In case II, the evolution of velocity profiles is also characterized by monotonic decreasing of $w(z, \tau_i\ (i = 1, \ldots, 5))$ with changing τ, before getting the equilibrium distribution $w_{eq}(z)$ across the LC channel. In this case, as shown in Figure 3b, the set of convex profiles, the vertical component of velocity $w(z, \tau_1)$ is directed in the positive direction. This distribution is characterized by the local maximum near the center of the LC sample $w(z = 0.46, \tau_5) = w_{eq}(z = 0.46) = 0.0009$ (0.016 μm/s), and the vertical component of the velocity is directed in the positive direction. A simple comparison of the maximum values of $w_{eq}(z = 0.3) = 0.01$ (0.2 m/s), in case I, and $w(z = 0.46, \tau_5) = w_{eq}(z = 0.46) = 0.0009$ (0.016 m/s), in case II, shows that in the first case, this value is almost an order of magnitude greater than in the second case.

Finally, the evolution of $w - u$ diagram for the case of lower cooler ($\chi_{z=0} = \chi_1$) and upper hotter ($\chi_{z=1} = \chi_1 + \Delta\chi$) bounding surfaces (case I), and for the number of dimensionless times $\tau_i = \tau_0 i$, $(i = 1, \ldots, 5)$, whose values increase from curve (1) to curve (5), is shown in Figure 4a.

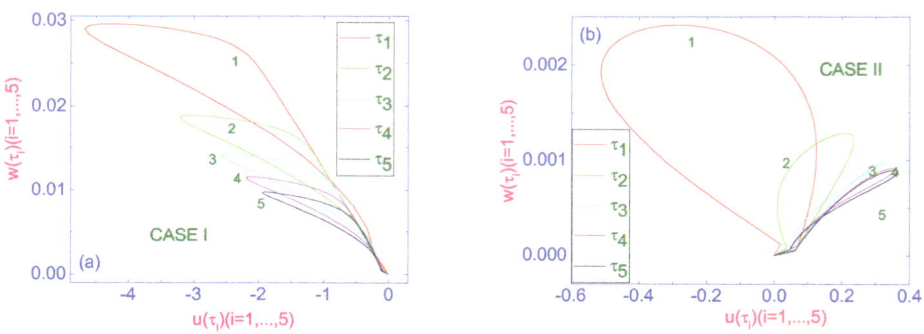

Figure 4. (**a**) Evolution of $w - u$ diagram for the case of lower cooler ($\chi_{z=0} = \chi_1$) and upper hotter ($\chi_{z=1} = \chi_1 + \Delta\chi$) bounding surfaces (case I), and for the number of dimensionless times $\tau_i = \tau_0 i$, $(i = 1, \ldots, 5)$, whose values increase from curve (1) to curve (5). Here, $\chi_1 = 0.97$, $\Delta\chi = 0.0162$ and $\tau_0 = 0.1$, respectively. (**b**) Same as in (**a**), but for case II.

Here, $\chi_1 = 0.97$, $\Delta\chi = 0.0162$ and $\tau_0 = 0.1$, respectively. This behavior is characterized by a monotonic decrease in the absolute values of both velocity components u and w, with the vertical component of the velocity directed in the positive direction, while the horizontal component is directed in the negative direction. When the LC sample is heated from below (case II), the evolution of $w - u$ diagram for the number of dimensionless times $\tau_i = \tau_0 i$, $(i = 1, \ldots, 5)$, whose values increase from curve (1) to curve (5), is shown in Figure 4b. In this case, the evolution of $w - u$ diagram is more complicated. During the first time period, up to $\tau_1 = 0.1$, the horizontal component of the velocity $u(\tau_1)$ changed direction from negative to positive, while during the subsequent time interval $\tau_5 - \tau_1 = 0.4$, the evolution of $w - u$ diagram is characterized by a monotonous increase in the values of u, and by simultaneously increasing the value of w.

It should be noted that in cases I and II, hydrodynamic flows of different magnitude and direction are formed in the LC channel under the influence of the same temperature gradient $\nabla\chi$.

The evolution of dimensionless shear stress tensor (ST) component $\sigma_{xz}(z, \tau_i)$ $(i = 1, \ldots, 5)$ as the function of the dimensionless distance z counted from the lower cooler $\chi_{z=0} = \chi_1$ to the upper warmer $\chi_{z=1} = \chi_1 + \Delta\chi$ (case I) bounding surface, during the first 5 dimensionless time terms $\tau_i = \tau_0 i$, $(i = 1, \ldots, 5)$, whose values increase from curve (1) to curve (5), is shown in Figure 5a.

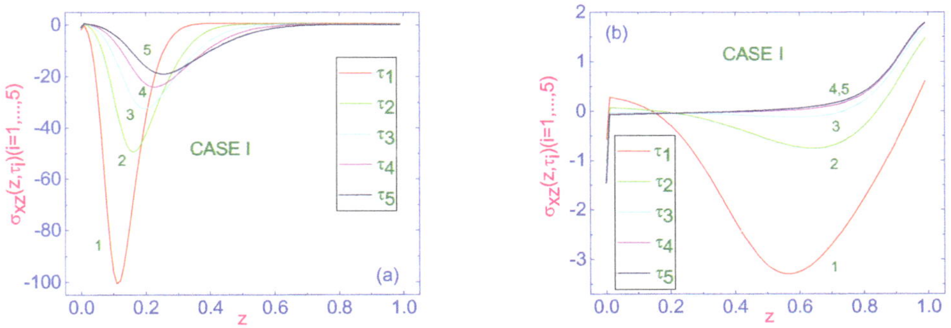

Figure 5. (**a**) Relaxation of dimensionless shear ST component $\sigma_{xz}(z, \tau_i)$ $(i = 1, \ldots, 5)$ as the function of the dimensionless distance z counted from the lower cooler $\chi_{z=0} = \chi_1$ to the upper warmer $\chi_{z=1} = \chi_1 + \Delta\chi$ (case I) bounding surface, during the first 5 dimensionless time terms $\tau_i = \tau_0 i$, $(i = 1, \ldots, 5)$, whose values increase from curve (1) to curve (5). Here, $\chi_1 = 0.97$, $\Delta\chi = 0.0162$ and $\tau_0 = 0.01$. (**b**) Same as in Figure 5a, but during the last 5 dimensionless time terms $\tau_i = \tau_0 i$, $(i = 1, \ldots, 5)$, where $\tau_0 = 0.1$.

Here, $\chi_1 = 0.97$, $\Delta\chi = 0.0162$ and $\tau_0 = 0.01$. During the first 5 time terms $\tau_i = \tau_0 i$, $(i = 1, \ldots, 5)$, the maximum absolute value of $|\sigma_{xz}(z, \tau)|$ decreases rapidly with $|\sigma_{xz}(z, \tau_1)| \sim 100$ to $|\sigma_{xz}(z, \tau_5)| \sim 20$, by about five times less, and this maximum is located near the lower cooler bounding surface. One of the features of this relaxation is that the values of the shear ST tensor $\sigma_{xz}(z, \tau)$ are negative during the first 5 time terms $\tau_i = \tau_0 i$, $(i = 1, \ldots, 5)$. During the last 5 time terms, $\tau_i = \tau_0 i$, $(i = 1, \ldots, 5)$ (see Figure 5b), where $\tau_0 = 0.1$, the value of $\sigma_{xz}(z, \tau_1)$ decreases rapidly near the lower bounding surface from $\sigma_{xz}(z \sim 0, \tau_1) \sim -1.5$ to $\sigma_{xz}(z \sim 0, \tau_1) \sim 0.25$ with further oscillation across the thickness of the HACN channel. The equilibrium distribution of $\sigma_{xz}(z, \tau_5) \equiv \sigma_{xz}(z, \tau_{eq})$ is characterized by a monotonous increase in $\sigma_{xz}(z, \tau)$ from -1.5 on the lower bounding surface to 1.8 on the upper hotter bounding surface.

In turn, the evolution of dimensionless normal ST component $\sigma_{xx}(z, \tau_i)$ $(i = 1, \ldots, 5)$ as the function of the dimensionless distance z, during the first 5 dimensionless time terms $\tau_i = \tau_0 i$, $(i = 1, \ldots, 5)$, whose values increase from curve (1) to curve (5), is shown in Figure 6a.

As in the case of shear ST, during the first 5 time terms $\tau_i = \tau_0 i$, $(i = 1, \ldots, 5)$, the maximum absolute value of $|\sigma_{xx}(z, \tau)|$ decreases rapidly with $|\sigma_{xx}(z, \tau_1)| \sim 135$ to $|\sigma_{xx}(z, \tau_5)| \sim 20$, about more than 6 times less, and this maximum is located near the lower cooler bounding surface. One of the features of this relaxation is that the values of the normal ST tensor $\sigma_{xx}(z, \tau)$ are negative during the first 5 time terms $\tau_i = \tau_0 i$, $(i = 1, \ldots, 5)$, over most of the channel thickness, while in the area close to the lower restricted surface, these values are positive. During the last 5 time terms, $\tau_i = \tau_0 i$, $(i = 1, \ldots, 5)$ (see Figure 6b), where $\tau_0 = 0.1$, the value of $\sigma_{xx}(z, \tau_1)$ is characterized by a monotonous increase from the negative $\sigma_{xx}(z \sim 0, \tau_1) \sim -2.0$, near the lower surface, to $\sigma_{xx}(z \sim 1.0, \tau_5) \sim 4.0$, near the upper surface, respectively.

The relaxation of the rest ST components $\sigma_{zx}(\tau)$ and $\sigma_{zz}(\tau)$ to their equilibrium values $\sigma_{zx}^{eq}(\tau)$ and $\sigma_{zz}^{eq}(\tau)$, during the last 5 time terms, $\tau_i = \tau_0 i$, $(i = 1, \ldots, 5)$, where $\tau_0 = 0.1$, and described by Equations (22) and (23), are shown in Figure 7a and 7b, respectively.

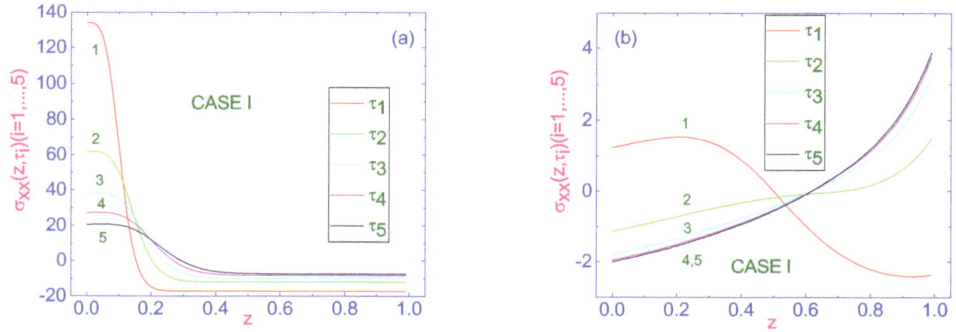

Figure 6. Same as in Figure 5a, but for the dimensionless normal ST component $\sigma_{xx}(z,\tau_i)$ $(i = 1,\ldots,5)$ as the function of the dimensionless distance z and time τ_i. (**a**) For the first 5 dimensionless time terms $\tau_i = \tau_0 i$, $(i = 1,\ldots,5)$, with $\tau_0 = 0.01$, while (**b**) is for the last 5 dimensionless time terms $\tau_i = \tau_0 i$, $(i = 1,\ldots,5)$, where $\tau_0 = 0.1$, respectively.

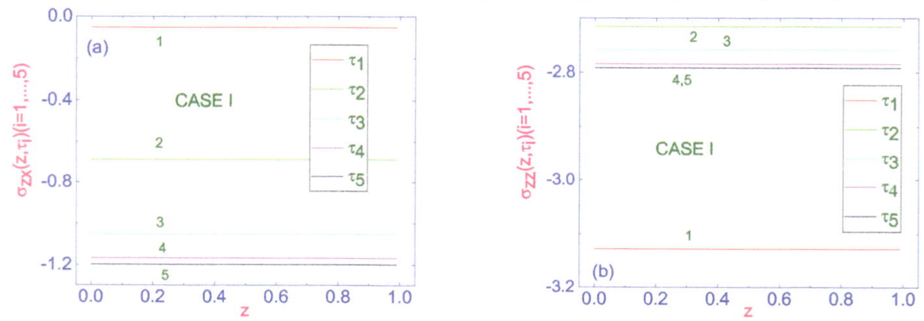

Figure 7. Same as in Figure 5a, but for the dimensionless ST components $\sigma_{zx}(\tau_i)$ $(i = 1,\ldots,5)$ (**a**) and $\sigma_{zz}(\tau_i)$ $(i = 1,\ldots,5)$ (**b**) as the function of the dimensionless distance z and time τ_i, respectively. In both cases (**a**,**b**), the results are given for the last 5 dimensionless time terms $\tau_i = \tau_0 i$, $(i = 1,\ldots,5)$, with $\tau_0 = 0.1$.

Taking into account that both ST components $\sigma_{zx}(\tau) = C_1(\tau)$ and $\sigma_{zz}(\tau) = C_2(\tau)$ do not depend on z, they have been fixed by the boundary conditions. Both the equilibrium distributions of $\sigma_{zx}(\tau_5) \equiv \sigma_{zx}(\tau_{eq}) = -1.2$ and $\sigma_{zz}(\tau_5) \equiv \sigma_{zz}(\tau_{eq}) \sim -2.8$ are characterized by negative values. The above distributions of the ST components show that in the HACN channel, under the effect of the temperature gradient $\nabla\chi$, directed from cooler lower to hotter upper bounding surfaces, a complex, mainly horizontal flow is excited in the negative direction (see Figure 3a).

In case II, when the LC sample is heated from below with the dimensionless temperature difference $\Delta\chi = 0.0162$ (~ 5 K), the evolution of dimensionless shear ST component $\sigma_{xz}(z,\tau_i)$ $(i = 1,\ldots,5)$ as the function of the dimensionless distance z counted from the lower hotter $\chi_{z=0} = \chi_1 + \Delta\chi$ to the upper cooler $\chi_{z=1} = \chi_1$ bounding surface, during the first 5 dimensionless time terms $\tau_i = \tau_0 i$, $(i = 1,\ldots,5)$, whose values increase from curve (1) to curve (5), is shown in Figure 8a.

Here, $\chi_1 = 0.97$, $\Delta\chi = 0.0162$ and $\tau_0 = 0.01$. During the first 5 time terms $\tau_i = \tau_0 i$, $(i = 1,\ldots,5)$, the maximum absolute value of $|\sigma_{xz}(z,\tau)|$ decreases rapidly with $|\sigma_{xz}(z,\tau_1)| \sim 15$ to $|\sigma_{xz}(z,\tau_5)| \sim 2.0$, by about 7 times less, and this maximum is located near the center of the HACN channel. One of the features of this relaxation is that the values of shear ST tensor $\sigma_{xz}(z,\tau)$ are negative during the first 5 time terms $\tau_i = \tau_0 i$, $(i = 1,\ldots,5)$. During the last 5 time terms, $\tau_i = \tau_0 i$, $(i = 1,\ldots,5)$ (see Figure 8b), where $\tau_0 = 0.1$, the value of $\sigma_{xz}(z,\tau_1)$ decreases rapidly near the lower bounding surface from $\sigma_{xz}(z \sim 0,\tau_1) \sim 1.2$ to $\sigma_{xz}(z \sim 0,\tau_1) \sim -0.25$, with a further decrease to the value $\sigma_{xz}(z \sim 0.5,\tau_1) \sim -2.7$, and a further gradual increase up to 0.8, near the upper bounding surface. Equilibrium distribution of $\sigma_{xz}(z,\tau_5) \equiv \sigma_{xz}(z,\tau_{eq})$ is characterized by an almost constant small negative value across the LC channel.

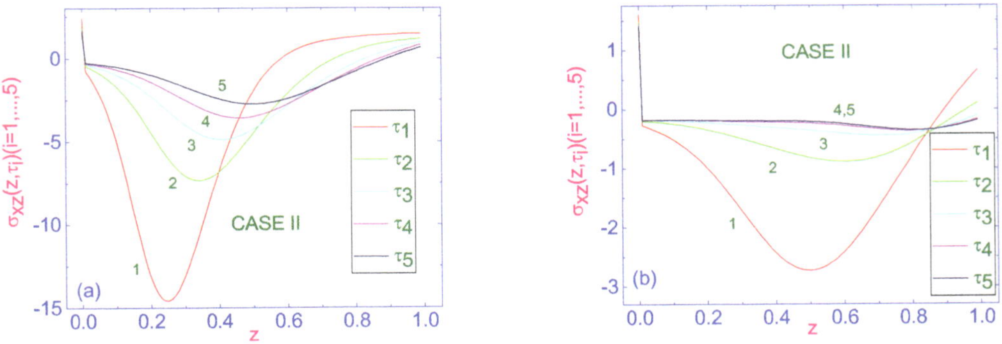

Figure 8. For both cases (**a**,**b**), the results are given for the same conditions as in Figure 5a, but for case II.

In turn, the evolution of dimensionless normal ST component $\sigma_{xx}(z,\tau_i)$ $(i = 1,\ldots,5)$ as the function of the dimensionless distance z, during the first 5 time terms $\tau_i = \tau_0 i$, $(i = 1,\ldots,5)$, whose values increase from curve (1) to curve (5), is shown in Figure 9a.

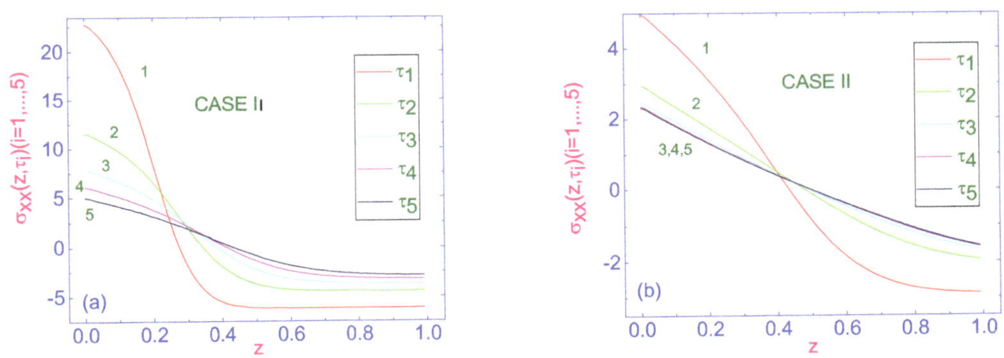

Figure 9. Same as in Figure 6a, but for case II. (**a**) For the first 5 dimensionless time terms $\tau_i = \tau_0 i$, $(i = 1,\ldots,5)$, with $\tau_0 = 0.01$, while (**b**) is for the last 5 dimensionless time terms $\tau_i = \tau_0 i$, $(i = 1,\ldots,5)$, where $\tau_0 = 0.1$, respectively.

As in case I, during the first 5 time terms $\tau_i = \tau_0 i$, $(i = 1,\ldots,5)$, the maximum absolute value of $|\sigma_{xx}(z,\tau)|$ decreases rapidly with $|\sigma_{xx}(z,\tau_1)| \sim 23$ to $|\sigma_{xx}(z,\tau_5)| \sim 5$, by about more than 4 times less, and this maximum is located near the lower hotter bounding surface. One of the features of this relaxation is that the values of the normal ST component $\sigma_{xx}(z,\tau)$ are positive during the first 5 time terms $\tau_i = \tau_0 i$, $(i = 1,\ldots,5)$, only near the lower hotter bounding surface, while in the rest area of the LC channel, these values are negative. During the last 5 time terms, $\tau_i = \tau_0 i$, $(i = 1,\ldots,5)$ (see Figure 9b), where $\tau_0 = 0.1$, the value of $\sigma_{xx}(z,\tau_5)$ is characterized by a gradual decrease from the positive $\sigma_{xx}(z \sim 0,\tau_5) \sim 2.3$, near the lower surface, to $\sigma_{xx}(z \sim 1.0,\tau_5) \sim -1.6$, near the upper surface, respectively.

The relaxation of the rest ST components $\sigma_{zx}(\tau)$ and $\sigma_{zz}(\tau)$ to their equilibrium values $\sigma_{zx}^{eq}(\tau)$ and $\sigma_{zz}^{eq}(\tau)$, during the last 5 time terms, $\tau_i = \tau_0 i$, $(i = 1, \ldots, 5)$, where $\tau_0 = 0.1$, and described by Equations (22) and (23), are shown in Figure 10a and 10b, respectively.

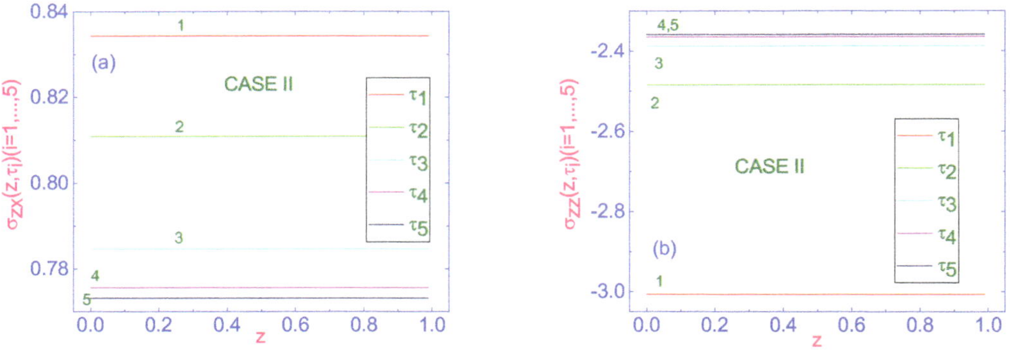

Figure 10. For both cases (**a**,**b**), the results are given for the same conditions as in Figure 7a, but for case II.

As in case I, in case II, both ST components $\sigma_{zx}(\tau) = \mathcal{C}_1(\tau)$ and $\sigma_{zz}(\tau) = \mathcal{C}_2(\tau)$ do not depend on z, and they have been fixed by the boundary conditions. Both the equilibrium distributions of $\sigma_{zx}(\tau_5) \equiv \sigma_{zx}(\tau_{eq}) = 0.773$ and $\sigma_{zz}(\tau_5) \equiv \sigma_{zz}(\tau_{eq}) \sim -2.45$ are characterized by positive and negative values, respectively. The above distributions of the ST components show that in the HACN channel, under the effect of the temperature gradient $\nabla\chi$, directed from cooler upper to hotter lower bounding surfaces, a complex, mainly horizontal flow is excited in the positive direction (see Figure 3b).

It should be noted that in cases I and II, hydrodynamic flows of different magnitudes and directions are formed in the LC channel under the effect of the same temperature gradient $\nabla\chi$ but directed in the opposite directions.

4. Conclusions

In summary, we have investigated the relaxation of director $\hat{\mathbf{n}}(\mathbf{r}, t)$, velocity $\mathbf{v}(\mathbf{r}, t)$, and stress tensor components $\sigma_{ij}(z, t)$ (i, j = x, z) in the HACN channel to their equilibrium values, under the influence of the temperature gradient directed normal to the bounding surfaces. Our calculations, based upon the classical Leslie–Ericksen theory, show that the hybrid-aligned compressible nematic material under the effect of the temperature gradient ∇T, directed from the lower cooler to upper hotter bounding surfaces, settles down to a stationary flow regime with the horizontal u and vertical w components of velocity, which sufficiently differ from these components under the effect of the same ∇T but directed from lower hotter to upper cooler bounding surfaces.

Note that in the case of a hybrid-aligned incompressible nematic (HAIN) channel, where due to the incompressibility condition $\nabla \cdot \mathbf{v} = 0$ and the boundary no-slip condition for the velocity field there is only one nonzero component of the vector $\mathbf{v}$, *viz.*, $\mathbf{v} = v_x\hat{\mathbf{i}} = u\hat{\mathbf{i}}$, the character of the stationary distribution of the horizontal component of velocity $u_{eq}(z)$ (HAIN) differs sufficiently from the same stationary distribution of the horizontal component of velocity $u_{eq}(z)$ (HACN) for a hybrid-aligned compressible nematic channel. This difference is due to the different behavior of $\nabla\hat{\mathbf{n}}$ and $\nabla\chi$ across the LC sample.

We believe that the present study may shed some light on the problem of the reorientation process in a compressible nematic channel under the influence of a vertical temperature gradient. We also believe that this paper shows not only some useful routes for estimating relaxation times but also for analyzing the remaining problems associated with an LC′ device's stability, efficiency, and lifetime.

Author Contributions: Software, I.Ś. and P.V.M.; validation, I.Ś. and A.V.Z.; investigation, I.Ś., P.V.M. and A.V.Z.; writing—review and editing, I.Ś.; supervision, A.V.Z. All authors have read and agreed to the published version of the manuscript.

Funding: A.V. Z. acknowledges financial support of the Ministry of Science and Higher Education of the Russian Federation for IPMash RAS (Grant Number FFNF-2024-0009).

Data Availability Statement: The original contributions presented in this study are included in the article; further inquiries can be directed to the corresponding author.

Conflicts of Interest: The authors declare no conflicts of interest.

Appendix A. Torques and Stress Tensor Components

The dimensionless density is $\rho \equiv \rho(z.\tau) = \rho(z,\tau)/\rho_0$, where ρ_0 is the mass density of the LC phase, $\rho_{,\tau} = \partial\rho(z,\tau)/\partial\tau$, $\tau = (K_{10}/\gamma_{10}d^2)t$ is the dimensionless time, and $\overline{w} = \frac{\gamma_{10}d^2}{K_{10}}w$ is the vertical dimensionless component of velocity, while K_{10} and γ_{10} are the highest values of the splay constant $K_1(\chi)$ and RVC $\gamma_1(\chi)$ in the temperature interval $[\chi_1, \chi_2]$ belonging to the nematic phase, respectively. Here, $\chi(z,\tau) = T(z,\tau)/T_{\rm NI}$ is the dimensionless temperature, $T_{\rm NI}$ is the nematic–isotropic transition temperature, $\gamma_1(\chi)$ and $\gamma_2(\chi)$ are the temperature-dependent rotational viscosity coefficients (RVCs), $K_1(\chi)$ and $K_3(\chi)$ are the splay and bend elastic constants of the LC phase.

The torque balance equation can be derived from the dimension balance of elastic $\mathbf{T}_{\rm el} = \frac{\delta \mathcal{W}_{\rm el}}{\delta \hat{\mathbf{n}}} \times \hat{\mathbf{n}}$, viscous $\mathbf{T}_{\rm vis} = \frac{\delta \mathcal{R}^{\rm vis}}{\delta \hat{\mathbf{n}}_t} \times \hat{\mathbf{n}}$, and thermomechanical $\mathbf{T}_{\rm tm} = \frac{\delta \mathcal{R}^{\rm tm}}{\delta \hat{\mathbf{n}}_t} \times \hat{\mathbf{n}}$ torques, where $\mathcal{W}_{\rm el} = \frac{1}{2}\left[K_1(\nabla \cdot \hat{\mathbf{n}})^2 + K_3(\hat{\mathbf{n}} \times \nabla \times \hat{\mathbf{n}})^2\right]$ is the elastic energy, K_1 and K_3 are the splay and bend elastic coefficients, and $\hat{\mathbf{n}}_{\rm t} \equiv \frac{d\hat{\mathbf{n}}}{dt}$ is the material derivative of $\hat{\mathbf{n}} = n_x\hat{\mathbf{i}} + n_z\hat{\mathbf{k}}$.

In the case of compressible fluid, the dimensionless linear momentum balance equation reduces to [10]

$$\delta_3\rho\frac{dw(z,\tau)}{d\tau} = \sigma_{zz,z}, \tag{A1}$$

$$\delta_3\rho\frac{du(z,\tau)}{d\tau} = \sigma_{zx,z}, \tag{A2}$$

where $\sigma_{zi} = \sigma^v_{zi} + \sigma^e_{zi}$ $(i = x,z)$ are both the shear and normal ST components, $\sigma^v_{zx}(\tau) = \frac{\delta\mathcal{R}(\tau)}{\delta u_{,z}}$ and $\sigma^v_{zz}(\tau) = \frac{\delta\mathcal{R}(\tau)}{\delta w_{,z}}$ are the viscous contributions to these ST components, $\mathcal{R}(z,\tau) = \frac{\gamma_{10}d^4}{K_{10}^2}\mathcal{R}(z,t)$ is the full dimensionless Rayleigh dissipation function, where $\mathcal{R}(z,t) = \mathcal{R}_{vis} + \mathcal{R}_{tm} + \mathcal{R}_{th}$, and $\mathcal{R}_{vis} = \frac{1}{2}h(\theta)u^2_{,z} - \overline{\mathcal{A}}(\theta)\theta_t u_{,z} + \frac{1}{2}\gamma_1\theta^2_{,t} + \frac{\beta_0}{6}u_{,z}w_{,z}\sin 2\theta + (\frac{\beta_0}{18} + \frac{1}{2}\xi_v)w^2_{,z}$ is the viscous, $\mathcal{R}_{tm} = \xi\theta_{,t}\theta_{,z}T_{,z}\left(\frac{1}{2} + \sin^2\theta\right) - \frac{3}{2}\xi T_{,z}u_{,z}\theta_{,z}\sin^2\theta - \frac{\xi}{4}T_{,z}w_{,z}\theta_{,z}$ $\sin 2\theta$ is the thermomechanical, and $\mathcal{R}_{th} = \frac{1}{2T}\left(\lambda_{\|}\cos^2\theta + \lambda_{\perp}\sin^2\theta\right)T^2_{,z}$ is the thermal contributions, respectively, whereas $\sigma^e_{zz} = -\overline{\mathcal{G}}(\theta)\theta^2_{,z} = -2W_F$, and $\sigma^e_{zx} = 0$ are the elastic contributions to the normal and shear components of the stress tensor. Here $h(\theta) = \alpha_1(T)\sin^2\theta\cos^2\theta - \overline{\mathcal{A}}(\theta)\theta_{,t}u_{,z} + \frac{1}{2}\alpha_4(T) + g(\theta)$, $g(\theta) = \frac{1}{2}(\alpha_6(T)\sin^2\theta + \alpha_5(T)\cos^2\theta)$, $T_{,z} = \partial T(z,t)/\partial z$, $\beta_0 = \alpha_1 + \alpha_5 + \alpha_6 + \frac{3}{2}\alpha_4$, $\delta_3 = \rho_0K_{10}/\gamma^2_{10}$ is a parameter of the system, and ξ_v is the volume viscosity, while $\alpha_i(T)$ $(i = 1,\ldots,6)$ are the temperature-dependent six Leslie coefficients, $\lambda_{\|}$ and $\lambda_{\perp}$ are the heat conductivity coefficients parallel and perpendicular to the director $\hat{\mathbf{n}}$. Note that in the case of a very thin LC layer, the hydrostatic pressure gradient ∇P contributions to both Equations (A1) and (A2) are equal to zero.

In the case of compressible nematic fluid $\partial_t\rho + \nabla\cdot(\rho\mathbf{v}) = 0$, it is convenient to rewrite the Rayleigh dissipation function $\mathcal{R}(z,t)$ in the form with and without accounting for the compressible term as [19]

$$\mathcal{R}(z,t) = \mathcal{R}_1(z,t) + \mathcal{R}_2(z,t), \tag{A3}$$

where $\mathcal{R}_1(z,t) = \left(\text{œ} + \frac{\partial W_F}{\partial\nabla\hat{\mathbf{n}}}\cdot\hat{\mathbf{n}}\nabla\right):\mathbf{v}\nabla = \sigma^v:\mathbf{v}\nabla$ and $\mathcal{R}_2(z,t) = \left(\text{œ}^m - \hat{\mathbf{n}}\times\frac{\partial W_F}{\partial\hat{\mathbf{n}}\nabla}\right):\vec{\omega}\nabla - \vec{\omega}\cdot\mathcal{E}:\left(\sigma + \frac{\partial W_F}{\partial\nabla\hat{\mathbf{n}}}\cdot\hat{\mathbf{n}}\nabla\right) = \sigma^{mv}:\vec{\omega}\nabla - \vec{\omega}\cdot\mathcal{E}:\sigma^v$, respectively. Here, $\mathbf{v}$ and $\vec{\omega}$ are the linear and angular velocities, $\nabla\mathbf{v}$ and $\nabla\vec{\omega}$ are the gradients of these fields, $\text{œ} = \sigma^v + \sigma^e$ and $\text{œ}^m = \sigma^{mv} + \sigma^{me}$ are the stress and the momentum stress tensors, $\mathcal{E}$ is the Levi-Civita tensor, $\sigma^v = \frac{\partial\mathcal{R}}{\partial\nabla\mathbf{v}}$, $\sigma^e = -\frac{\partial W_F}{\partial\nabla\hat{\mathbf{n}}}\cdot\hat{\mathbf{n}}\nabla$, $\sigma^{mv} = \frac{\partial\mathcal{R}}{\partial\nabla\vec{\omega}}$, and $\sigma^{me} = \hat{\mathbf{n}}\times\frac{\partial W_F}{\partial\hat{\mathbf{n}}\nabla}$ are the viscous and elastic contributions to the total stress tensor σ and momentum ST σ^m, respectively. We use the invariant, multiple dot convention: $\mathbf{a}\cdot\mathbf{b} = a_ib_i$, $\mathbf{A}\cdot\mathbf{B} = A_{ik}B_{kj}$, and $\mathbf{A}:\mathbf{B} = A_{ik}B_{ki}$, where repeated Cartesian indices are summed. By means of decomposition of the tensors on the spherical, symmetric, and asymmetric deviators, one can rewrite the last two tensors and gradients $\mathbf{v}\nabla$ and $\vec{\omega}\nabla$ in the form [19] $\sigma^v = \sigma_0^v\mathcal{I} + \sigma_s^v + \sigma_a^v$, $\sigma^{mv} = \sigma_0^{mv}\mathcal{I} + \sigma_s^{mv} + \sigma_a^{mv}$, $\mathbf{v}\nabla = v_0\mathcal{I} + \mathbf{v}\nabla_s + \mathbf{v}\nabla_a$, and $\vec{\omega}\nabla = \omega_0\mathcal{I} + \vec{\omega}\nabla_s + \vec{\omega}\nabla_a$, respectively. Here, $\mathcal{I}$ is the unit tensor. Now the full dissipation function $\mathcal{R}$ can be written as

$$\begin{aligned}\mathcal{R}(z,t) &= \sigma_0^v v_0\mathcal{I}:\mathcal{I} + \sigma_s^v:\mathbf{v}\nabla_s + \sigma_a^v:\mathbf{v}\nabla_a \\ &+ \sigma_0^{mv}\omega_0\mathcal{I}:\mathcal{I} + \sigma_s^{mv}:\vec{\omega}\nabla_s + \sigma_a^{mv}:\vec{\omega}\nabla_a.\end{aligned} \tag{A4}$$

In the following, one deals with the complicated flow $\mathbf{v} = (u,0,w)$ excited by the temperature gradient and the director reorientation with four components of the viscous stress tensor $\sigma_{xx}^v, \sigma_{xz}^v, \sigma_{zx}^v$, and σ_{zz}^v, which are connected to each other by relations $2\sigma_0^v = \sigma_{xx}^v + \sigma_{zz}^v$, $2v_0 = w_{,z}$, and $2\sigma_0^v = \partial\mathcal{R}/\partial v_0$, respectively. Taking into account Equations (A3) and (A4), the equations for the tensor components σ_{xx}^v and σ_{zz}^v can be written in the form

$$\sigma_{zz}^v + \sigma_{xx}^v = 2\frac{\partial\mathcal{R}(z,t)}{\partial w_{,z}}, \tag{A5}$$

$$\sigma_{zz}^v = \frac{\partial\mathcal{R}(z,t)}{\partial w_{,z}}. \tag{A6}$$

The last system of Equations (A5) and (A6) has a solution

$$\sigma_{zz}^v = \sigma_{xx}^v.$$

Note that the compressible part of $\mathcal{R}(z,t)$ can be decomposed as $\frac{1}{2}P\nabla\cdot\mathbf{v} + \frac{1}{2}\sigma_s^v:(\nabla\mathbf{v} + \mathbf{v}\nabla) + \frac{1}{2}\sigma_a^v:(\nabla\mathbf{v} - \mathbf{v}\nabla)$, where σ_s^v and σ_a^v are symmetric and asymmetric deviators of the tensor σ^v, whereas $P = \sigma^v:\mathcal{I} = \sigma_{xx}^v + \sigma_{zz}^v$ is the viscous contribution to the total pressure $\overline{P}$. The elastic contribution to $\overline{P}$ is equal to $P^e = \sigma^e:\mathcal{I} = \sigma_{xx}^e + \sigma_{zz}^e = -2W_F$. So, the total dimensionless pressure is given by $\overline{P} = 2(\sigma_{zz}^v - W_F)$.

References

1. de Gennes, P.G.; Prost, J. *The Physics of Liquid Crystals*, 2nd ed.; Oxford University Press: Oxford, UK, 1995.
2. Whitesides, G.M. The origins and the future of microfluidics. *Nature* **2006**, *442*, 368–373. [CrossRef] [PubMed]
3. Whitby, M.; Quirke, N. Fluid flow in carbon nanotubes and nanopipes. *Nat. Nanotechnol.* **2007**, *2*, 87–94. [CrossRef]
4. Rey, A.D. Liquid crystal models of biological materials and processes. *Soft Matter* **2010**, *6*, 3402–3429. [CrossRef]
5. Cuennet, J. G.; Vasdekis, A. E.; De Sio, L.; Psalis, D. Optofluidic modulator based on peristaltic nematogen microflows. *Nat. Photon.* **2011**, *5*, 234–238. [CrossRef]
6. Woltman, S.J.; Jay, G.D.; Crawford, G.P. Liquid-crystal materials find a new order in biomedical applications. *Nat. Mater.* **2007**, *6*, 929–938. [CrossRef]
7. Schoch, R.B.; Han, J.; Renaud, P. Transport phenomena in nanofluidics. *Rev. Mod. Phys.* **2008**, *80*, 839–883. [CrossRef]
8. Squires T.M.; Quake, S.R. Microfluidics: Fluid physics at the nanoliter scale. *Rev. Mod. Phys.* **2005**, *77*, 977–1026. [CrossRef]

9. Nazari, M.; Davoodabadi, A.; Huang, D.; Luo, T.; Ghasemi, H. Transport Phenomena in Nano/Molecular Confinements. *ACS Nano* **2020**, *14*, 16348–16391. [CrossRef] [PubMed]
10. Śliwa, I.; Zakharov, A.V. Heat driven flow in microsized nematic volumes: Computational studies and analysis. *Symmetry* **2021**, *13*, 459. [CrossRef]
11. Śliwa, I.; Maslennikov, P.V.; Zakharov, A.V. Vortex Dynamics in a Hybrid Aligned Nematic Microvolume with an Orientational Defect. *Symmetry* **2023**, *15*, 324. [CrossRef]
12. Yamamoto, K.; Ota, N.; Tanaka, Y. Nanofluidic Devices and Applications for Biological Analyses. *Anal. Chem.* **2021**, *93*, 332–349. [CrossRef]
13. Ericksen, J.L. Anisotropic Fluids. *Arch. Ration. Mech. Anal.* **1960**, *4*, 231–237. [CrossRef]
14. Leslie, F.M. Some constitutive equations for liquid crystals. *Arch. Ration. Mech. Anal.* **1968**, *28*, 265–283. [CrossRef]
15. Landau, L.D.; Lifshitz, E.M. *Fluid Mechanics*; Pergamon Press: Oxford, UK, 1987.
16. Cross, M.C.; Hohenberg, P.C. Pattern formation outside of equilibrium. *Rev. Mod. Phys.* **1993**, *65*, 851–1111. [CrossRef]
17. Graham, R. Hydrodynamic fluctuations near the convection instability. *Phys. Rev. A* **1974**, *10*, 1762–1784. [CrossRef]
18. Berezin, I.S.; Zhidkov, N.P. *Computing Methods*, 4th ed.; Clarendon: Oxford, UK, 1965.
19. Stewart, I.W. *The Static and Dynamic Continuum Theory of Liquid Crystals*; Taylor and Francis: London, UK, 2004.

crystals

MDPI

Article

Light-Induced Higher-Order Bragg Resonance in Heliconical Cholesteric Liquid Crystals

Ashot H. Gevorgyan [1] and Francesco Simoni [2,3,*]

[1] School of Natural Sciences, Far Eastern Federal University, 10 Ajax Bay, Russky Island, 690922 Vladivostok, Russia; agevorgyan@ysu.am

[2] Università Politecnica delle Marche, 60131 Ancona, Italy

[3] Institute of Applied Sciences and Intelligent Systems CNR, 80072 Pozzuoli, Italy

* Correspondence: f.simoni@photomat.it

Abstract: Higher-order reflections from a cholesteric liquid crystal have been investigated for a long time after early observations of an oblique incidence of light. Even with the restriction of the required oblique geometry, wavelength tuning would make this property very attractive for photonic applications in the near-UV range of the electromagnetic spectrum. Unfortunately, this is quite difficult to achieve in cholesterics, and it has been obtained in special configurations that lead to the deformation of the helical structure. We show here that a new opportunity is provided by heliconical cholesteric liquid crystals and demonstrate that second-order Bragg resonance with high reflectivity and a narrow band can be produced by a light beam with intensity high enough to give rise to a significant optical torque with the consequent distortion of the helical structure. Exploiting the well-known properties of heliconical cholesterics under electric fields, we also demonstrate that the easy tuning of the second-order reflection band is achieved in the near-UV range.

Keywords: higher-order Bragg resonance; heliconical liquid crystals; nonlinear optics; pitch tuning

1. Introduction

Second-order Bragg reflection from a cholesteric liquid crystal (CLC) was observed by Berreman and Scheffer in the first experiments on single-domain CLC samples for the oblique incidence of light [1,2]. This peculiar optical property of CLCs becomes attractive for several reasons: the possibility of generating a reflection band in the blue or near-UV range of the electromagnetic spectrum, the provision of a reflection band much narrower than the first-order Bragg resonance, and a much faster response time [3,4]. Of course, providing the additional property of the continuous electrical tuning of the higher-order reflection over a broad wavelength range makes this phenomenon more interesting for photonic applications. However, the extended investigations carried out on this subject have shown that the electrical tuning of CLC Bragg resonance is not easy to achieve. To overcome these difficulties, different configurations have been proposed.

Using CLC planar cells with interdigitated electrodes on the boundaries, second- and third-order reflection bands were observed over a critical value of the applied field [4,5]. The limit of this configuration is the need for a quite large voltage to be applied, the broad reflection spectrum, and the red-shifting due to the helix unwinding.

A different approach has been proposed by D.J. Broer and coworkers [6] using CLC polymers, where it is shown that deformation of the helical structure gives rise to higher-order reflections with an efficiency dependent on the strength of deformation. This result is

https://doi.org/10.3390/cryst15060513

quite important since it clarifies the origin of the appearance of the higher-order reflections due to the application of an electric field to the CLC structure.

A more recent paper by K.M. Lee et al. [7] reports the formation of higher-order reflections without the need for interdigitated electrodes or the involvement of polymerization process using polymer-stabilized CLCs (PSCLCs). In this case, the spectral position of the second-order reflection is determined by the concentration of chiral dopant and can be controlled by an applied electric field.

A breakthrough in the tunability of the Bragg diffraction over the entire visible spectrum has been represented in the past decade by the discovery of the "oblique helicoidal structure" (ChOH): a special configuration of CLC occurring when the condition $K_3 < K_2$ is fulfilled and an electric field in a range of values within two critical values ($E_{N^*C} < E_S < E_{NC}$) is applied along the helix direction. Under these conditions, the stable structure is given by the molecular director $\mathbf{n}$ rotating on a conical surface around the helical axis, thus forming the so-called "heliconical cholesteric liquid crystal" [8–12]. (Conventional definitions for the bend elastic constant K_3 and the twist elastic constant K_2 are used above).

Due to the presence of a director tilt angle θ with respect to the helix axis in the conical structure, easy modulation of the helix pitch becomes possible in ChOH, thus allowing for tuning the Bragg resonance over the whole visible spectrum by scanning a low-frequency electric field over fractions of V/µm [8–10]. This property makes this configuration extremely interesting for developing tunable photonic devices, overcoming a major obstacle present in the usual planar configuration of CLCs. For this reason, it also stimulates the investigation focused on conditions allowing for the generation of higher-order reflections. A recent report highlights many unexpected features of reflections originated by ChOH for the oblique incidence of the incoming light, analyzing (in much detail) the reflection bands originated by the pitch p of the helical structure and those originated by the fractions $p/2$, $p/3$, and $p/4$, with spectral location determined by the incidence angle and the applied voltage [13,14].

On the other hand, the bend deformation present in ChOH allows for efficient coupling between the optical field and the molecular director and gives rise to light-induced optical reorientation when light intensity is high enough to provide an effective optical torque [15]. This effect can produce strong nonlinear behavior in the propagation of the exciting light beam [16–18]. The modification of the transmission spectrum occurring during light propagation leads to self-induced oscillations on the edges of the reflection band and the broadening of the Bragg resonance, which becomes unstable and disappears at high intensities, underlying a strong light-induced deformation of the helical structure [19]. Additionally, the inclusion of photosensitive compounds in the material composition has been demonstrated to be effective in controlling the reflection band in ChOH [20], thus providing another method to induce a nonlinear light propagation.

As a consequence, taking into account the results mentioned above concerning reflection bands in CLCs and ChOH, we should expect the appearance of higher-order reflection peaks in ChOH even for normal incidences of light when nonlinear optical reorientation is effective.

In this paper, we show that distortion of the helical structure induced by optical reorientation leads to higher-order narrow reflection peaks for light normally impinging on a ChOH sample. The second-order Bragg resonance gives rise to significant and stable reflectivity (>0.5). Additionally, broad wavelength tuning is possible in the near-UV range, providing new features to develop tunable UV filters for photonic applications.

2. Theoretical Approach and Method

We give here a brief summary of the main steps of the approach for calculating the optical properties for a light beam travelling through a ChOH sample in the presence of an effective torque induced by the optical field on the molecular director. Details can be found in previous publications [17–19].

The starting point is the dielectric permittivity matrix at an optical frequency of ChOH, written as

$$\mathbf{M}^{ChOH} = \begin{bmatrix} \widetilde{\epsilon} + \frac{\Delta\epsilon_{eff}}{2} cos(2qz) & \frac{\Delta\epsilon_{eff}}{2} sin\,(2qz) \\ \frac{\Delta\epsilon_{eff}}{2} sin(2qz) & \widetilde{\epsilon} - \frac{\Delta\epsilon_{eff}}{2} cos\,(2qz) \end{bmatrix} \quad (1)$$

Here, $\Delta\epsilon_{eff} = \epsilon_{eff} - \epsilon_{\perp}$, $\widetilde{\epsilon} = \frac{1}{2}\left(\epsilon_{eff} + \epsilon_{\perp}\right)$, $\epsilon_{eff} = \frac{\epsilon_{\perp}\epsilon_{\parallel}}{\epsilon_{\parallel} - (\epsilon_{\parallel} - \epsilon_{\perp}) sin^2\theta}$, where, as usual, $\epsilon_{\parallel}$ and $\epsilon_{\perp}$ are the dielectric permittivities parallel and perpendicular to the director in the local frame. The sample considered is formed of a large number of sublayers (identified by the number, m), each of them with constant optical parameters, and the calculation is based on Ambartsumian's layer-addition modified method, which allows for the recalculation of the optical field after traveling through each sublayer based on the field propagation equations.

This method provides the reflection coefficient $R = \frac{|E_r|^2}{|E_i|^2}$, the transmission coefficient $T = \frac{|E_t|^2}{|E_i|^2}$, and the light intensity of the wave travelling in the ChOH sample $I(z) = |E_{in}(z)|^2 I_i$. Here $E_{in}(z)$ is the total wave field arising in the ChOH layer at the z point. The method of this calculation in all detail is described in [21].

The calculation has been performed both for the diffracting eigenmode with the circular polarization with the same handedness of the ChOH helix, and for the non-diffracting mode with circular polarization with opposite handedness. The light is impinging at normal incidence, where intensity $I_i = NI_0$. In this way, we use the normalized intensity N, considering, in the first step of the calculation ($j = 1$), the effective applied field to be [15–19].

$$E_{eff} = \sqrt{E_S^2 - \frac{\Delta\epsilon}{\Delta\epsilon_S} N} = \text{const} \quad (2)$$

with the helix pitch $p = \text{const}$ and $s = sin^2\theta = \text{const}$.

Here, E_S is the low-frequency (static) field necessary to stabilize the heliconical structure, $\Delta\epsilon = \epsilon_{\parallel} - \epsilon_{\perp}$ is the dielectric anisotropy at optical frequencies, and $\Delta\epsilon_S$ is the value at low frequencies.

The calculation allows obtaining $I(z,\lambda) = |E_{in}(z,\lambda)|^2 I_i$ to be used in the next steps to provide the new effective field $E_{eff}(z,\lambda) = \sqrt{E_S^2 - \frac{\Delta\epsilon_{OPT}}{\Delta\epsilon} |E_{in}(z,\lambda)|^2 N}$. This allows us to obtain the new angle and pitch distribution:

$$s(z,\lambda,j) = sin^2\theta = \frac{\kappa}{1-\kappa}\left(\frac{E_{NC}}{E_{eff(z,\lambda,j)}} - 1\right) \text{and } p(z,\lambda,j) = \kappa \frac{E_{NC} p_0}{E_{eff(z,\lambda,j)}} \quad (3)$$

Thus, the reflectivity and transmittivity of the ChOH sample are obtained for each impinging intensity.

The material parameters chosen for the calculation are the same reported in ref. [10] and used in our former calculations. They refer to a multi-component composition made by a mixture of flexible dimers M (Merck), a mixture of rod-like mesogens E7 (Merck), heptane CB7CB, and a chiral dopant S811 (Merck) with the following weight proportion M:E7:CB7CB:S811 = 45.5:30.3:20.2:4.0. Details on the material are reported in [10]. Then, we have $\varepsilon_{\parallel} = 2.79$, $\varepsilon_{\perp} = 2.19$, $\Delta\varepsilon = 4.79$, $\kappa = 0.1$, and the unperturbated helix pitch is

p_0 = 1400 nm, with critical fields $E_{NC} = 4.88$ V/μm and $E_{N^*C} = 1.53$ V/μm. The static field applied along the helix direction is $E_S = 2.03$ V/μm. In our calculations, all the above-mentioned sublayers have the same thickness: $d_1 = d_2 = d_3 = \ldots = d_L = 8$ nm.

We remark that computer programs were compiled in Visual Basic and debugged by the authors, and the graphs were built using the Excel 16.89 and MATLAB R2022a programs.

3. Results and Discussion

In Figure 1, the reflection spectra for a 20 μm thick sample are reported for two values of the impinging intensity. At $N = 1$, we observe the usual Bragg resonance with a peak located near 500 nm and a bandwidth of about 10 nm. At $N = 2000$, the Bragg spectrum is reduced and suffers a slight blue shift and large broadening to more than 50 nm, while a second peak with significant reflectivity at the second harmonic wavelength of the original Bragg peak appears. This phenomenon is evident in the corresponding Figure 2, where the spectrum evolution is reported versus the parameter j, counting the steps of the calculation that measures the elapsed time t (using the response time τ_0 of the material as the unit [19]). The spectrum calculated at $N = 1000$ shows that as soon as the main resonance broadens due to the nonlinear optical reorientation effect, additional higher-order resonances appear. In the following, we focus on the second-order resonance, because the increasing UV absorption at shorter wavelengths may affect the actual experimental observation of the higher-order reflections, depending on the chemical structure of the mixture components [22,23].

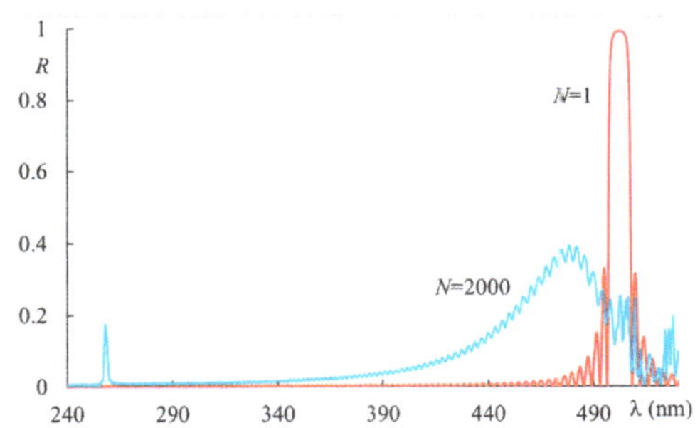

Figure 1. Reflectivity vs. wavelength for a sample thickness $d = 20$ μm, applied static field $E_S = 2.03 \frac{\text{V}}{\mu\text{m}}$, with material parameters indicated in the text, and two different values of the light intensity: red curve $N = 1$, blue curve $N = 2000$. The step index $j = 2$ (see text).

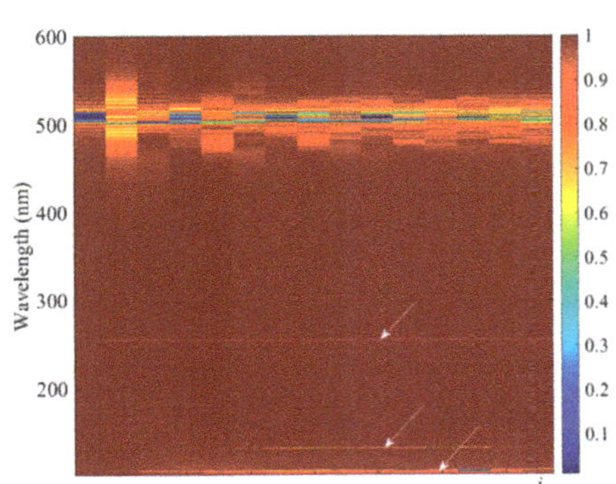

Figure 2. The transmission spectra vs. the step-index j. Colors show the normalized value of the transmittivity. The index j is proportional to the elapsed time (see text). Light intensity $N = 1000$, and the other parameters are as in Figure 1. Arrows indicate the presence of higher-order resonances.

Driven by this result, we have studied the main features of this phenomenon. In Figure 3, we report the reflectivity of the second-order Bragg diffraction for increasing values of the impinging light intensity. We see that it is barely present at $N = 500$, reaching a significant 0.2 value at $N = 2000$ and over 0.6 at $N = 5000$, even if this last value is not stable (as we will show later). A very important feature of this resonance is the bandwidth being much narrower than the first-order Bragg resonance: about 2 nm. It is quite interesting to understand the characteristics of the process, and the results are shown in Figure 4, where the second-order resonance is reported at $N = 1000$ for increasing values of the step parameter j. We observe that the maximum value is achieved for $j = 2$, while relaxation and stabilization for a lower value of reflectivity occur for higher values of j, i.e., on a longer time scale.

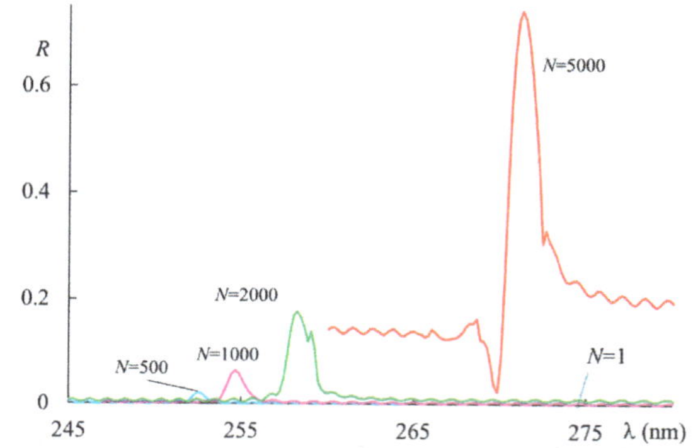

Figure 3. Reflectivity vs. wavelength of the second-order Bragg resonance for a sample thickness $d = 20$ μm for increasing values of impinging light intensity. The applied static field $E_S = 2.03$ V/μm, with material parameters indicated in the text. The step index $j = 2$ (see text).

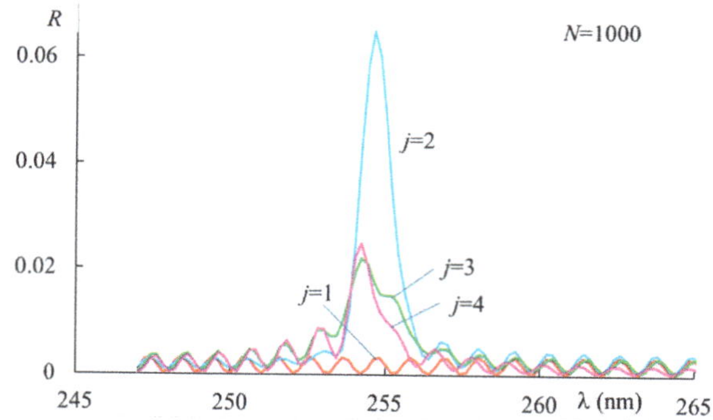

Figure 4. Reflectivity vs. wavelength of the second-order Bragg resonance for a sample thickness $d = 20$ μm at $N = 1000$ for increased step-index j. The index j is proportional to the elapsed time (see text). The applied static field $E_S = 2.03$ V/μm, with material parameters indicated in the text.

Quite different is the behavior at a somewhat high intensity ($N = 5000$), as shown in Figure 5. In this case, we have a well-defined peak and high reflectivity for $j = 2$ (over 0.60), but it disappears for higher values of j because the system becomes unstable and the spoiling of the Bragg resonance occurs, as already reported [19].

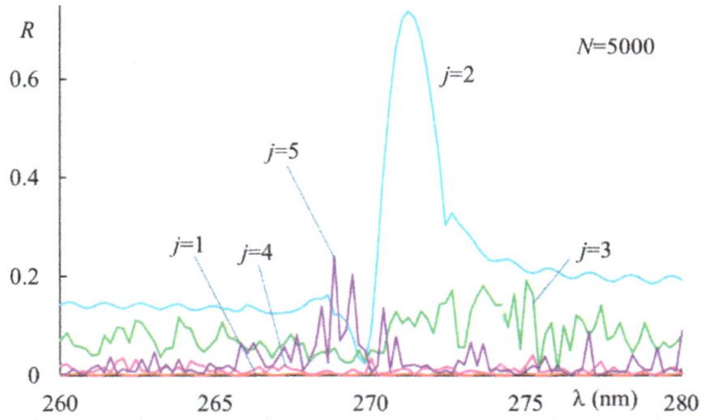

Figure 5. Reflectivity vs. wavelength of the second-order Bragg resonance for a sample thickness $d = 20$ μm at $N = 5000$ for increased step-index j. The index j is proportional to the elapsed time (see text). The applied static field $E_S = 2.03$ V/μm, with material parameters indicated in the text.

We have also checked the effect of sample thickness on the reflectivity of the second-order resonance. We chose a value for the light intensity ($N = 2000$) that leads to a significant reflectivity while allowing for the stabilization of the structure. These data are reported in Figure 6 for three different values of the sample thickness. They show an increase in reflectivity accompanied by an increased background around the peak.

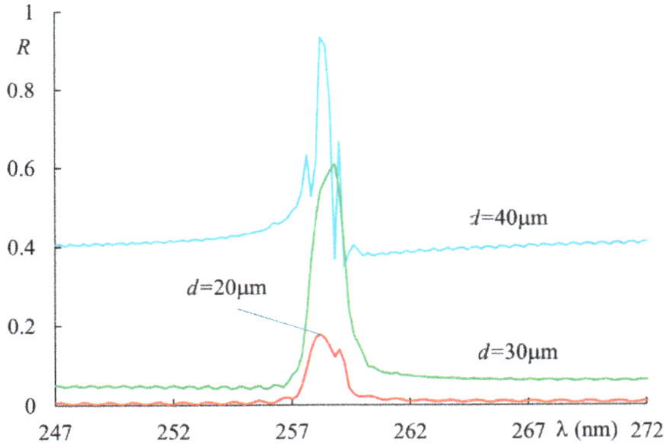

Figure 6. Reflectivity vs. wavelength of the second-order Bragg resonance for increasing values of the sample thickness at $N = 2000$. The applied static field $E_S = 2.03$ V/μm, with material parameters indicated in the text.

The presented results are easily explained, taking into account the effect of the induced optical reorientation on the propagation of the light beam. We have previously demonstrated oscillations and the spoiling of the Bragg resonance induced by light in ChOH. This is due to a transition from a stable conical structure to a disordered system occurring by increasing the light intensity to a level high enough to produce a significant optical torque on the molecular director. We have shown that the shift and broadening of the Bragg resonance is accompanied by side-band oscillations, corresponding to the strong distortion of the helical structure, still identified by a central peak. On the other hand, we remind readers in the introduction that higher-order Bragg reflections can be originated by an inhomogeneously twisted director in a CLC structure [6]. Therefore, the observed second-order Bragg reflection in the figures reported is understood as having originated from the helix distortion. The time evolution of the spectrum at the high intensity reported in Figure 5 strengthens this explanation. In fact, we obtain high reflectivity on a short time scale ($j = 2$) when the helix is still defined, even if quite distorted, while reflectivity drops and is completely spoiled on a longer time scale because the helix cannot be identified any more.

Based on these results, it is possible to exploit the peculiar properties of ChOH related to the easy tuning of the Bragg resonance over a broad wavelength range to obtain a narrow peak in the near-UV range that is continuously tuned by varying the applied static field. To achieve this goal, we chose a light intensity and sample thickness that led to a distorted but stable helical structure. Additionally, we considered the initial pitch of the correspondent CLC helix $p_0 = 1700$ nm in order to produce the second-order resonance in the near-UV range. (The pitch can be easily adjusted by controlling the concentration of the chiral dopant in the components of the material). Figure 7 shows the reflectivity under the chosen conditions for low ($N = 1$) and high intensities ($N = 2000$), as in Figure 1. We see the first-order Bragg resonance near 600 nm and a clear second-order narrow peak near 300 nm.

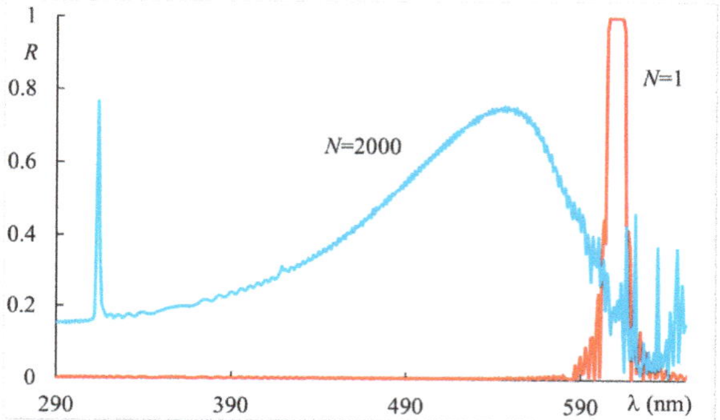

Figure 7. Reflectivity vs. wavelength for a sample thickness $d = 42.5$ μm, applied static field $E_S = 2.03$ V/μm, $p_0 = 1700$ nm, with material parameters indicated in the text and two different values of the light intensity: red curve $N = 1$, blue curve $N = 2000$. The step index $j = 2$ (see text).

This narrow and high reflection band is easily tuned in the near-UV range by a small variation of the applied static field, as shown in Figure 8.

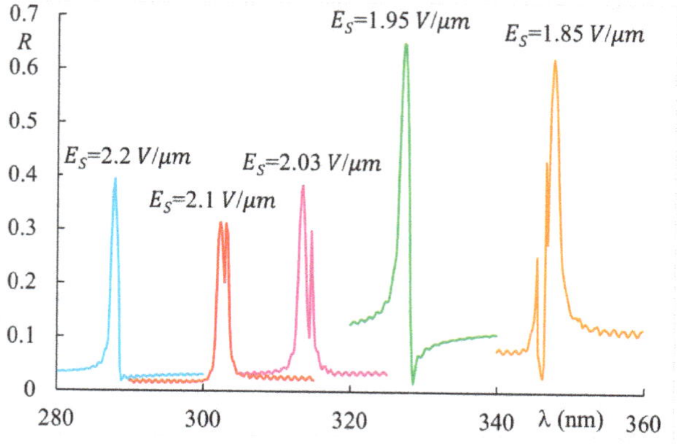

Figure 8. Reflectivity vs. wavelength of the second-order Bragg resonance for a sample thickness $d = 34$ μm, $p_0 = 1700$ nm, with material parameters indicated in the text. The impinging light intensity is $N = 2000$. The step index $j = 2$ (see text).

Here, we show wavelength tuning in the range 350–290 nm of a narrow peak (bandwidth 2 nm) with a reflectivity around 0.4 over a variable background. The effective tuning range can be made broader depending on the material parameters and the light absorption at short wavelengths. This result demonstrates the possibility of achieving tunable and high-selectivity filtering in the near-UV range and is quite attractive for photonic applications.

4. Conclusions

We have shown that heliconical cholesteric liquid crystals represent a new opportunity for the investigation and possible exploitation of higher-order reflections originated by the helical structure. We have demonstrated that second-order Bragg resonance with a reflectivity of 0.4–0.5 and bandwidth of 2 nm is originated by the distortion of the helical structure—a consequence of optical reorientation produced by a light beam impinging at normal incidence on the sample. The variation of the applied static electric field allows for the easy tuning of the second-order reflection band in the near-UV range.

Author Contributions: Conceptualization A.H.G. and F.S.; theoretical approach F.S.; numerical calculations A.H.G.; data analysis A.H.G. and F.S.; writing—review and editing, A.H.G. and F.S. All authors have read and agreed to the published version of the manuscript.

Funding: A.H.G. acknowledge the financial support from the Russian Science Foundation (Grant No. 24-22-00283).

Data Availability Statement: The data presented in this study are available on request from the author A.H.G. due to privacy reason.

Conflicts of Interest: The authors declare no conflicts of interest.

References

1. Berreman, D.W.; Scheffer, T.J. Bragg Reflection of Light from Single-Domain Cholesteric Liquid-Crystal Films. *Phys. Rev. Lett.* **1970**, *25*, 577–581. [CrossRef]
2. Berreman, D.W.; Scheffer, T.J. Bragg Reflection and Transmission by Single-Domain Cholesteric Liquid-Crystal Films: Theory and Verification. *Mol. Cryst. Liq. Cryst.* **1970**, *11*, 395–405. [CrossRef]
3. Blinov, L.M. *Structure and Properties of Liquid Crystals*; Springer: Berlin/Heidelberg, Germany, 2011.
4. Blinov, L.M.; Belyaev, S.V.; Kizel, V.A. High-order reflections from a cholesteric helix induced by an electric field. *Phys. Lett.* **1978**, *A65*, 33–35. [CrossRef]
5. Rumi, M.; White, T.J.; Bunning, T.J. Reflection spectra of distorted cholesteric liquid crystal structure in cells with interdigitated electrodes. *Opt. Exp.* **2014**, *22*, 16510–16519. [CrossRef]
6. Broer, D.J.; Mol, G.N.; van Haaren, J.A.M.M.; Lub, J. Photo-Induced Diffusion in Polymerizing Chiral-Nematic Media. *Adv. Mater.* **1999**, *11*, 573–578. [CrossRef]
7. Lee, K.M.; Reshetnyak, V.Y.; McConney, M.E.; Crenshaw, E.P.; Bunning, T.J.; White, T.J.; Godman, N.P. Higher-Order Bragg Reflection Colors in Polymer-Stabilized Cholesteric Liquid Crystals. *Adv. Photonics Res.* **2021**, *2*, 2100112. [CrossRef]
8. Xiang, J.; Shiyanovskii, S.V.; Imrie, C.T.; Lavrentovich, O.D. Electrooptic response of chiral nematic liquid crystals with oblique heliconical director. *Phys. Rev. Lett.* **2014**, *112*, 217801. [CrossRef]
9. Xiang, J.; Li, Y.; Li, Q.; Paterson, D.A.; Storey, J.M.D.; Imrie, C.T.; Lavrentovich, O.D. Electrically tunable selective reflection of light from ultraviolet to visible and infrared by heliconical cholesterics. *Adv. Mater.* **2015**, *27*, 3014. [CrossRef]
10. Mrukiewicz, M.; Iadlovska, O.S.; Babakhanova, G.; Siemianowski, S.; Shiyanovskii, S.V.; Lavrentovich, O.D. Wide temperature range of an electrically tunable selective reflection of light by oblique helicoidal cholesteric. *Liq. Cryst.* **2019**, *46*, 1544. [CrossRef]
11. De Gennes, P.G. Calcul de la distorsion d'une structure cholesterique par un champ magnetique. *Solid State Commun.* **1968**, *6*, 163. [CrossRef]
12. Meier, R.B. Effects of electric and magnetic fields on the structure of cholesteric liquid crystals. *Appl. Phys. Lett.* **1968**, *12*, 281. [CrossRef]
13. Mrukiewicz, M.; Iadlovska, O.S.; Shiyanovskii, S.V.; Lavrentovich, O.D. Diffraction of obliquely incident light at oblique helicoidal cholesteric. In *Liquid Crystals XXVI*; SPIE: Bellingham, WA, USA, 2022; Volume 12207, pp. 63–68.
14. Iadlovska, O.S.; Thapa, K.; Rajabi, M.; Mrukiewicz, M.; Shiyanovskii, S.V.; Lavrentovich, O.D. Electrically tunable total reflection of light by oblique helicoidal cholesteric. *MRS Bull.* **2024**, *49*, 835–850. [CrossRef]

15. Nava, G.; Ciciulla, F.; Iadlovska, O.S.; Lavrentovich, O.D.; Simoni, F.; Lucchetti, L. Pitch tuning induced by optical torque in heliconical cholesteric liquid crystals. *Phys. Rev. Res.* **2019**, *1*, 033215. [CrossRef]
16. Simoni, F. Nonlinear optical propagation near the Bragg resonance in heliconical cholesteric liquid crystals. *Opt. Lett.* **2020**, *45*, 6510. [CrossRef]
17. Gevorgyan, A.H.; Simoni, F. Onset of Optical Instabilities in the Nonlinear Optical Transmission of Heliconical Cholesteric Liquid Crystals. *Photonics* **2022**, *9*, 139–148. [CrossRef]
18. Gevorgyan, A.H.; Simoni, F. Nonlinear Optical Propagation in Heliconical Cholesteric Liquid Crystals. *Nov. Opt. Mater.* **2024**, *4*, 93–119.
19. Gevorgyan, A.H.; Simoni, F. Light-induced Self-oscillations and Spoiling of the Bragg Resonance due to Nonlinear Optical Propagation in Heliconical Cholesteric Liquid Crystals. *Photonics* **2022**, *9*, 881–892. [CrossRef]
20. Mrukiewicz, M.; Cigl, M.; Perkowski, P.; Karcz, J.; Hamplova, V.; Bubnov, A. Dual tunability of selective reflection by light and electric field for self-organizing materials. *J. Mol. Liq.* **2024**, *400*, 124540. [CrossRef]
21. Gevorgyan, A.H.; Harutyunyan, M.Z. Chiral photonic crystals with an anisotropic defect layer. *Phys. Rev. E* **2007**, *76*, 031701-9. [CrossRef]
22. Wu, S.-T.; Ramos, E.; Finkenzeller, U. Polarized UV spectroscopy of conjugated liquid crystals. *J. Appl. Phys.* **1990**, *78*, 85. [CrossRef]
23. Lin, P.-T.; Wu, S.-T.; Chang, C.-Y.; Hsu, C.-S. UV Stability of High Birefringence Liquid Crystals. *Mol. Cryst. Liq. Cryst.* **2004**, *411*, 243–253. [CrossRef]

crystals

MDPI

Article

Broadband Near-Infrared Reflective Film from Stacked Opposite-Handed Chiral Liquid Crystals with Pitch Gradients

Hyeon Seong Hwang [1,†], Jongsu Lee [2,†], Byungsoo Kang [3], Minhye Kim [1], Doyo Kim [1] and Se-Um Kim [1,*]

1 Department of Electrical and Information Engineering, Seoul National University of Science and Technology, Seoul 01811, Republic of Korea; he9312@seoultech.ac.kr (H.S.H.); minhyeral14@seoultech.ac.kr (M.K.); doyoo0413@seoultech.ac.kr (D.K.)
2 Sensor System Research Center, Korea Institute of Science and Technology, Seoul 02792, Republic of Korea; yukgagsu@kist.re.kr
3 Advanced Photovoltaics Research Center, Korea Institute of Science and Technology, Seoul 02792, Republic of Korea; bskang@kist.re.kr
* Correspondence: seumkim@seoultech.ac.kr
† These authors contributed equally to this work.

Abstract

Broadband near-infrared (NIR) reflective films are widely used in architecture and the automotive and aerospace industries for energy saving and thermal regulation. For large-area and flexible applications, it is essential to develop cost-effective, solution-processable, and long-term-stable NIR-reflective films. Here, we present a polymer-stabilized chiral liquid crystal (CLC) film that achieves broadband NIR reflection by stacking opposite-handed CLC layers with pitch gradients. We experimentally established optimal formulations of materials for both right-handed and left-handed CLCs. The resulting film exhibits high-degree broadband reflection (~95%) in the 1000–1800 nm wavelength range, while maintaining visible transmittance (~80%) in the 450–850 nm range. The concept proposed here will be widely applicable for scalable and practical NIR-filtering applications in smart glasses, sensors, and optoelectronic devices.

Keywords: near-infrared reflection; chiral liquid crystal; pitch gradient; polymer-stabilized liquid crystal; ultraviolet absorption dye

1. Introduction

A significant portion of solar energy, over 50%, lies in the near-infrared (NIR) spectrum, which contributes to excessive thermal load. This increases the cooling demand in buildings and, in the case of optoelectronic devices, accelerates performance degradation in outdoor environments. Therefore, the development of optical coatings or films that selectively reflect NIR wavelengths while maintaining high transmittance in the visible range is critical for improving thermal management performance in various applications [1–3].

Low-emissivity (low-e) coatings [4,5] and distributed Bragg reflectors (DBRs) [6], which are designed to strongly reflect in the NIR wavelength while remaining transparent in the visible range, have been widely employed to mitigate photothermal effects caused by NIR radiation. Low-e coatings incorporate thin, low-emissivity metal layers such as silver [4,5], aluminum [5,7], or copper [8,9] to reflect near-infrared (NIR) light, with metal oxide layers enhancing visible transmittance and protecting the metal layers. Their reflection typically spans from the NIR to IR region with high reflectance (over 90%), but the presence of metal layers often leads to reduced visible transmittance (typically 60–80%) [10].

DBRs, on the other hand, form a photonic bandgap by alternating high- and low- refractive-index dielectric layers to create destructive interference that blocks the propagation of NIR wavelengths [11–13]. DBRs generally offer a high visible transmittance of 80–90%. However, their NIR reflection bandwidth is relatively narrow (typically 100–200 nm) so that broadband photothermal effects remain constrained [14,15].

Despite their widespread use, the optical performance of both low-e coatings and DBRs strongly depends on the precise thickness of individual layers. For low-e coatings, thicker metal layers significantly reduce visible transmittance, while those thinner than a critical limit may result in incomplete conformal coverage on substrates [4,16]. Similarly, in DBRs, even slight deviations in the thickness of sub-layers shift the central reflection wavelength and reduce its intensity [17–19]. Moreover, achieving broadband reflection requires a substantial increase in the number of stacked layers [20,21]. These thin layers are typically fabricated via vacuum-assisted evaporation processes, which inherently limit scalability for large-area applications.

Cholesteric liquid crystals (CLCs) serve as polarization-dependent photonic crystals that enable thickness-independent NIR reflection owing to the helical structure of liquid crystal (LC) molecules. CLCs selectively reflect the circularly polarized light with the same handedness as the helical twist [22,23]. In CLCs, the central reflection wavelength λ and reflection bandwidth $\Delta\lambda$ are given by $\lambda = (n_e + n_o) \times P/2$ and $\Delta\lambda = (n_e - n_o) \times P$, where n_o and n_e are the ordinary and extraordinary refractive indices of the LC molecules, respectively, and P is the helical pitch. CLCs generally comprise host nematic liquid crystals doped with a small concentration of chiral compounds, and incorporating reactive monomers enables the formation of polymer-stabilized structures that can be fabricated into solid-state films [24–26]. These CLC films can be self-assembled through simple solution-based processes such as spin coating [27–30], spray coating [31,32], or bar-coating [33,34].

Since the reflection bandwidth of single-pitch CLCs is intrinsically narrow, various strategies have been proposed to broaden the reflection band by introducing a pitch gradient in CLCs through electrical modulation [25,35], thermal diffusion [36,37], or doping with ultraviolet (UV)-absorbing dyes [38–40]. CLCs are compatible with solution processing, large-area fabrication, and flexible substrates. However, their intrinsic polarization selectivity limits the maximum theoretical reflectance to 50%, as they only reflect circularly polarized light of a single handedness. To circumvent this, it is necessary to assemble CLC layers with opposite handedness into a monolithic structure capable of reflecting both left- and right-circularly polarized NIR light. Precise alignment of both the bandwidth and central wavelength of the reflection bands from the two CLC layers is essential to achieve almost-complete broadband NIR reflection. It is also important to maintain a high visible-light transmission in the assembled structure. Furthermore, the long-term reliability of such films under continuous exposure to intense NIR radiation is crucial for their practical implementation.

In this study, we demonstrate a broadband NIR-reflective film fabricated by stacking two opposite-handed CLC layers with pitch gradients. The pitch gradient was introduced by doping the CLCs with a UV-absorption dye, and the reflection characteristics were controlled by adjusting the concentrations of the dye and a reactive mesogen used for polymer stabilization. Each CLC layer was fabricated separately on different substrates and subsequently assembled into a single monolithic film. The resulting film exhibits high visible-light transmittance while achieving broad NIR reflection from 1000 to 1800 nm, with a maximum reflectance exceeding 95%. This combination of broadband NIR reflectance and high visible-light transmittance is comparable with those of established low-e and DBR-based films. The thermal durability of the film was evaluated, along with an assessment of its NIR reflection performance through infrared thermography to examine feasibility under practical conditions.

2. Materials and Methods

2.1. Materials

Reactive Mesogen RM257, host LC E7, and chiral dopant LC756 were purchased from Henan Wentao Chemical Product Co. (Zhengzhou, China). Chiral dopant S811 and UV-absorption dye UV-327 were purchased from Chemical Industry Co., Ltd. (Tokyo, Japan). Photoinitiator 2,2-Dimethoxy-2-phenylacetophenone (DMPA) was purchased from Sigma-Aldrich (St. Louis, MO, USA). The chemical structures of these materials are illustrated in Figure 1. Polyvinyl alcohol (PVA) and ethanol were purchased from Duksan General Science (Seoul, South Korea). Polyimide (PI) SE-7492 was provided by Nissan Chemical Co. (Tokyo, Japan) UV-curable adhesive NOA81 was purchased from Edmund Optics (Barrington, NJ, USA). All materials were used as received without further purification.

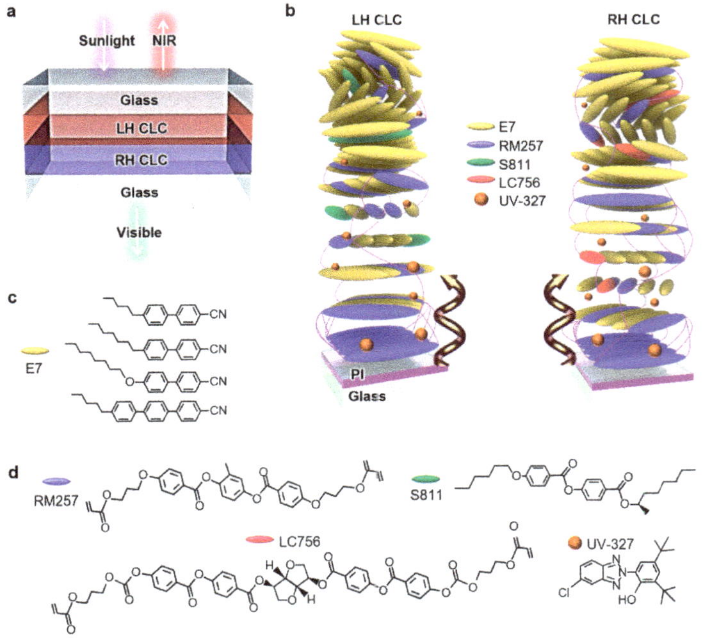

Figure 1. (**a**) Schematic illustration of the proposed NIR-reflective film structure. (**b**) Structural schematics of LH CLC (left) and RH CLC (right) layers. (**c**) Composition and chemical structures of the nematic LC mixture E7. (**d**) Chemical structures of the reactive mesogen RM257, the LH chiral dopant S811, the RH chiral dopant LC756, and the UV-absorption dye UV-327.

2.2. Sample Preparation

The CLC mixtures with various component ratios were prepared as described in Table 1. All sample groups contained 0.5 wt% of DMPA as the photoinitiator. The mixtures were stirred uniformly for 1 h at 100 °C and then slowly cooled to room temperature. Top and bottom glass substrates were cleaned with acetone and subsequently treated with UV-ozone for 5 min. After surface treatment, polyimide was spin-coated on the top substrate at a rate of 3000 rpm for 30 s and annealed at 230 °C for 2 h. A 1 wt% solution of PVA in distilled water was spin-coated on the bottom substrate at a rate of 3000 rpm for 30 s and annealed at 100 °C for 1 h. To achieve a homogeneous orientation of the CLC film, the

PI-coated top substrate and the PVA-coated bottom substrate were rubbed in anti-parallel directions and then assembled with a 25 μm cell gap, using 25 μm spacer particles dispersed in NOA81. The prepared CLC mixtures were inserted into the empty cell by capillary action at 35 °C, followed by 365 nm UV irradiation using a UIT250 UV source (SP-9, Ushio, Tokyo, Japan) with an output power of 1200 mW for 15 min to form a polymer network with a gradient pitch. The measurement of UV intensity (UV-340A, Lutron, Coopersburg, PA, USA) at the exposed surface was 2 mW/cm^2. The LC cells were then immersed in ethanol for 1 h to disassemble the substrates. The CLC layers were selectively retained on the top PI-coated substrate, which faced the UV light, due to stronger adhesion. Two types of CLC layers, left-handed (LH) CLC and right-handed (RH) CLC, were stacked with their CLC sides facing each other. In Table 1, the sample groups A and C are LH CLCs and the sample groups B and D are RH CLCs.

Table 1. Composition ratios of materials for CLC mixture preparation.

Sample	E7	RM257	S811	LC756	UV-327	DMPA
A1	73.5	15	10		1	0.5
A2	68.5	20	10		1	0.5
A3	63.5	25	10		1	0.5
A4	58.5	30	10		1	0.5
B1	81.5	15		2	1	0.5
B2	76.5	20		2	1	0.5
B3	71.5	25		2	1	0.5
B4	66.5	30		2	1	0.5
C1	59.5	30	10		0	0.5
C2	58.5	30	10		1	0.5
C3	57.5	30	10		2	0.5
C4	56.5	30	10		3	0.5
D1	67.5	30		2	0	0.5
D2	66.5	30		2	1	0.5
D3	65.5	30		2	2	0.5
D4	64.5	30		2	3	0.5

3. Results and Discussion

3.1. Structure and Chemical Composition of NIR-Reflective Films

The proposed NIR-reflective film is composed of two CLC layers with opposite handedness, an LH CLC layer and a RH CLC layer, as illustrated in Figure 1a. The helical pitches of these layers are designed to reflect NIR light while transmitting visible light. Figure 1b schematically shows the molecular arrangement of the LH CLC (left) and RH CLC (right). To achieve a chiral nematic phase at room temperature prior to polymerization, the nematic LC mixture, E7, was used as the host material. E7 consists of four mesogenic compounds: 4-cyano-4′-n-pentyl-biphenyl, 4-cyano-4′-heptyl-biphenyl, 4-cyano-4′-n-oxyoctyl-biphenyl, and 4-cyano-4″-n-pentyl-p-terphenyl (Figure 1c).

Figure 1d shows the chemical structures of the additional components: RM257, S811, LC756, and UV-327, which are dispersed in the host LC. The UV-assisted polymerization of the reactive mesogen RM257 stabilizes the overall LC structure. RM257 is a bifunctional linear reactive mesogen containing diacrylate groups, which forms the structural backbone of the CLC network and co-polymerizes with LC756. S811 and LC756 were used as LH and RH chiral dopants, respectively. It is important to note that S811 is unable to participate in the polymerization process, yet it induces a stabilized chiral nematic phase in the E7/RM257 composite during polymerization. As a result, the LH CLC layer retains its helical structure even after the removal of S811.

UV-327 was incorporated as a UV absorber [41,42] that modulates the UV light intensity throughout the film thickness, inducing a vertical gradient in the helical pitch [20,43]. This gradient arises from the diffusion behavior of monomers during polymerization, which depends on the local UV intensity. The presence of UV-absorption dye leads to reduced UV penetration, resulting in a higher monomer concentration near the UV-irradiated surface. This in turn lowers the relative concentration of chiral dopant, producing a longer helical pitch near the top substrate. Conversely, the helical pitch decreases near the bottom side of the film. This gradual variation in helical pitch across the film thickness produces a broadened reflection band in the NIR region [7,44,45].

3.2. Fabrication of the Optimal LH CLC and RH CLC Layers

An empty cell comprising two glass substrates, coated with thin layers of PI and PVA respectively, was prepared as shown in Figure 2a. A 25 μm cell gap was maintained using polystyrene (PS) spacer particles embedded at the four corners of the substrates. NOA81, serving as the dispersion medium for the spacer particles, was pre-cured under UV light to fix the gap. The prepared CLC mixtures were subsequently injected into the cell. After a second UV exposure to polymerize the injected mixtures, the sample was immersed in ethanol. After disassembling the cell, both LH (Figure 2b) and RH (Figure 2c) CLC layers were successfully fabricated on the PI-coated glass substrates. The PVA-coated glass substrate was easily detached without surface residues due to the weak adhesion between the CLC layer and the PVA surface [46–48]. The transmittance spectra of these layers are expected to exhibit full transmission in the visible region and polarization-dependent reflection of up to 50% in the NIR region.

The CLC layers can achieve near-theoretical maximum reflectance when the number of the helical pitches typically exceed 10 [49,50]. Based on the equation for the central reflection wavelength, $\lambda = (n_e + n_o) \times P/2$, and the approximate value $(n_e + n_o)/2 \sim 1.62$ given by the primary LC materials E7 [51] and RM257 [52], a 10 μm-thick layer can cover the maximum reflectance. However, bandwidth broadening was limited at this thickness (Figure 2d). A 25 μm-thick layer showed saturated bandwidth broadening without significant degradation in visible transmittance. Layers thicker than 25 μm may suffer reduced transmittance due to instability in LC alignment.

To target NIR reflection in the 1000–2000 nm range, the weight ratio of LC756 was fixed at 2 wt%. LC756 has a high helical twisting power (HTP) of 64 μm^{-1}, which is greater than that of S811 (11 μm^{-1}) [53,54]. Accordingly, the required S811 concentration to match the target reflection range is theoretically expected to be around 12 wt%. However, the actual matched concentration may vary depending on the host materials used [55,56]. To determine the optimal value, a series of samples was prepared with S811 concentrations ranging from 10 to 12 wt% in 1 wt% intervals. As shown in Figure 2e, the LH CLC layer containing 10 wt% S811 exhibited a reflection bandwidth well matched to that of the RH CLC layer, achieving broadband reflection in the 1000–1800 nm range. When the LH and RH CLC layers were stacked, the resulting NIR-reflective film demonstrated highly selective visible light transmission along with polarization-independent, significantly enhanced NIR reflectance (Figure 2e). Given the low optical absorption of CLC materials in the visible to NIR wavelength range [57], the reflectance spectrum (R) can be reliably inferred from the measured transmittance spectrum (T) as $R = 1 - T$. The stacked CLC layers exhibited approximately 80% visible transmittance at normal incidence. The remaining transmittance loss is primarily attributed to Fresnel reflections at the glass–air and internal layer interfaces, as well as intrinsic absorption from the glass substrate. No visible haze was observed even at oblique viewing angles (30° and 60°), confirming that the stacked CLC configuration achieved broadband NIR reflectance without introducing noticeable light scattering or haze in the visible range.

Although the reflectance spectrum shifts toward shorter wavelengths at oblique incidence, the broad NIR reflection bandwidth ensures that this spectral displacement remains within the effective operating range. Also, the reflection band was designed with sufficient angular tolerance by positioning its lower bound above 1000 nm so that the film remains fully transparent even at such oblique viewing directions.

The total reflectance of the stacked CLC configuration arises from the complementary polarization selectivity of the individual LH and RH CLC layers [58,59]. By combining opposite-handedness CLC layers, near-complete reflection of both circular polarization components is achieved, resulting in polarization-independent broadband NIR reflectance.

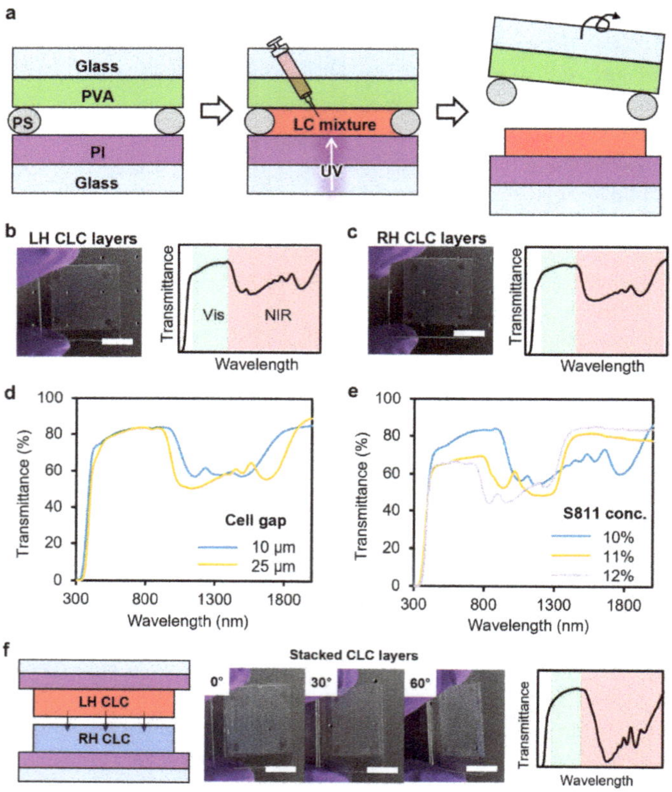

Figure 2. (**a**) Schematic illustration of the fabrication process for CLC layers. (**b**,**c**) Photographs of the fabricated (**b**) LH and (**c**) RH CLC layers with their corresponding transmittance spectra. (**d**) Transmittance spectra of LH CLC layers with 10 µm and 25 µm cell gap. (**e**) Transmittance spectra of LH CLC layers with varying concentrations of S811 (10, 11, and 12 wt%). (**f**) Schematic of the stacking process for LH and RH CLC layers, along with sample photographs taken at sample tilt angles of 0°, 30°, and 60°, and the representative transmittance spectrum. Scale bar, 1 cm.

3.3. Optimization of the Compositional Ratio and Fabrication Conditions of the CLC Mixture

To ensure the formation of a homogeneous polymer network capable of inducing a pitch gradient over a broad range in the NIR spectrum, it is essential to prevent phase separation between RM257-rich and E7-rich domains. Such phase separation leads to a decrease in visible transmittance due to haze, as well as distortion of the helical structure [60,61]. To determine the optimal weight ratio of RM257, its concentration was varied from 15 wt% to 30 wt% in 5 wt% increments for both LH CLC and RH CLC layers. As shown in the transmittance spectra for LH CLC (Figure 3a), increasing the RM257 concentration led to improved transmittance in the visible range and enhanced selective reflection in the NIR region. Polarized optical microscopy (POM) images reveal significant phase separation at lower RM257 concentrations of 15 wt% (Figure 3b) and 20 wt% (Figure 3c). At a concentration of 15 wt%, the domain size was smaller than the resolution limit of the microscope, appearing as subtle color fluctuations without clearly defined boundaries. In contrast, at 20 wt%, phase separation domains became physically distinguishable and were widely distributed across the sample.

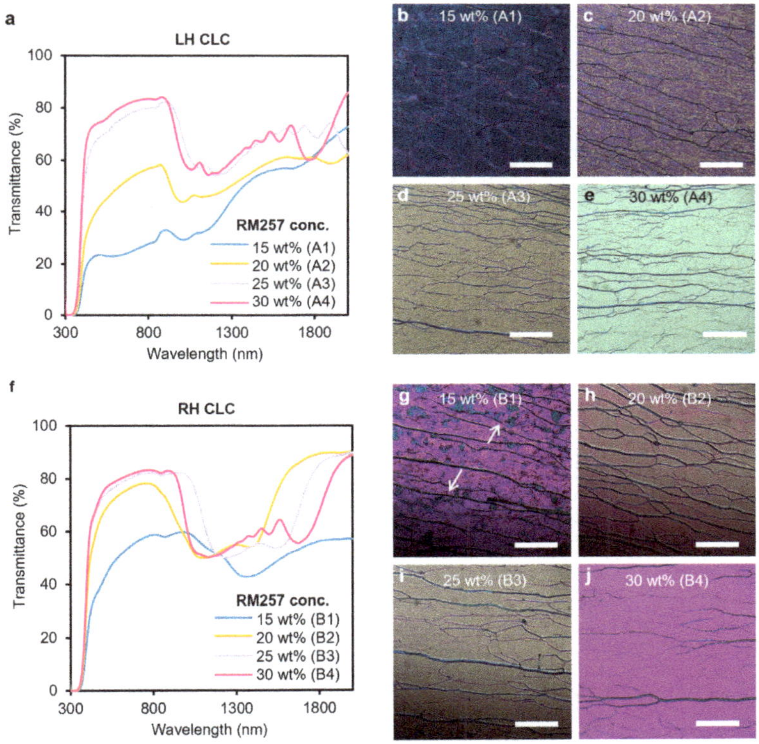

Figure 3. (**a**) Transmittance spectra of LH CLC layers with RM257 concentrations ranging from 15% to 30% (A1–A4). (**b–e**) POM images of LH CLC layers with RM257 concentrations of (**b**) 15%, (**c**) 20%, (**d**) 25%, and (**e**) 30%. (**f**) Transmittance spectra of RH CLC films with RM257 concentrations ranging from 15% to 30%. (**g–j**) POM images of RH CLC layers with RM257 concentrations of (**g**) 15%, (**h**) 20%, (**i**) 25%, and (**j**) 30%. In (**g**), the white arrows indicate representative regions of phase separation. Scale bar, 500 μm.

More uniform texture emerged at 25% RM257 (Figure 3d), with optimal uniformity and optical characteristics observed at 30% (Figure 3e). It should be noted that the oil streak textures observed in POM images were formed due to flow-induced alignment during LC injection [62,63].

For the RH CLC layer, a similar trend was observed. Increasing the RM257 concentration led to higher transmission in the visible range and broader NIR reflection (Figure 3f). At 15% RM257, a heterogeneous texture produced from phase separation was clearly visible (Figure 3g). In contrast to the LH CLC, a homogeneous phase appeared as early as 20% RM257 in the RH CLC, likely due to the copolymerization between RM257 and LC756 (Figure 3h). Both the uniformity and optical characteristics further improved at 25% (Figure 3i) and reached optimal levels at 30% RM257 (Figure 3j).

With a fixed RM257 concentration of 30 wt%, the UV-327 concentration was varied at 0%, 1%, 2%, and 3%, and the corresponding transmittance spectra were measured (Figure 4). For the LH CLC layer, visible-range transmittance decreased significantly when the UV-327 concentration reached 2% or higher (Figure 4a). Comparison of POM images at 0% (Figure 4b) and 1% (Figure 4c) UV-327 showed uniform phases in both cases. However, 1% UV-327 notably broadened the NIR reflection band by inducing a pitch gradient within the CLC structure. Compared with the undoped sample, the reflection bandwidth increased approximately four to five times. These results suggest that within the 0–1 wt% range, the pitch gradient magnitude increased approximately linearly with the UV-327 concentration [64]. However, higher UV-327 concentrations (2% and 3%) disrupted effective monomer polymerization and caused phase separation, as shown in Figure 4d and 4e, respectively.

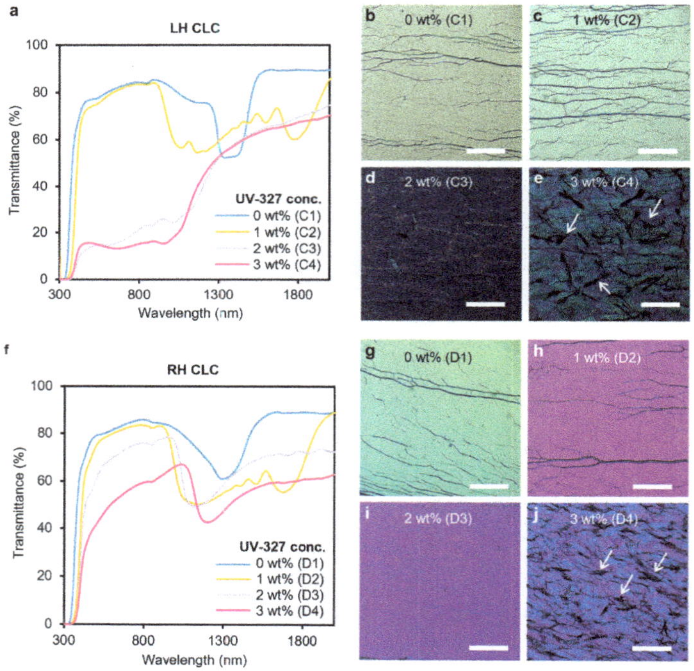

Figure 4. (**a**) Transmittance spectra of LH CLC layers with UV-327 concentrations ranging from 0% to 3%. (**b–e**) POM images of LH CLC layers with UV-327 concentrations of (**b**) 0%, (**c**) 1%, (**d**) 2%, and (**e**) 3%. (**f**) Transmittance spectra of RH CLC layers with UV-327 concentrations ranging from 0% to 3%. (**g–j**) POM images of RH CLC layers with UV-327 concentrations of (**g**) 0%, (**h**) 1%, (**i**) 2%, and (**j**) 3%. In (**e**,**j**), the white arrows indicate representative regions of phase separation. Scale bar, 500 μm.

For the RH CLC layer, visible transmittance gradually decreased, and the NIR reflection band narrowed as the UV-327 concentration increased (Figure 4f). As with the LH CLC layer, uniform LC phases were observed at both 0% (Figure 4g) and 1% UV-327 (Figure 4h), whereas phase separation occurred at 2% (Figure 4i) and 3% (Figure 4j). Therefore, a UV-327 concentration of 1% was identified as the optimal condition for both LH and RH CLC layers when combined with 30% RM257. It should be noted that in both LH and RH CLCs, an increase in the degree of phase separation with higher UV-327 concentrations appears as a growth in domain size observed in the POM images, particularly between 2% and 3% (Figure 4d,e,i,j).

With 30% RM257 and 1% UV-327 established as the optimal compositions for both LH and RH CLC layers, the effect of UV exposure time was further investigated. For the LH layer, the transmittance spectra exhibited a gradual broadening of the NIR reflection band and a slight decrease in visible transmission as the UV exposure time increased from 0 to 20 min (Figure 5a). However, considering sample-to-sample variations, the spectra appeared to saturate after approximately 10 min of exposure. This is supported by POM images, which show a clear difference between the unexposed sample and the one exposed for 10 min (Figure 5b,c). From 10 to 15 min of exposure, the LH CLC layer maintained its optical properties (Figure 5d), with a noticeable overall decrease in transmittance observed at 20 min (Figure 5e).

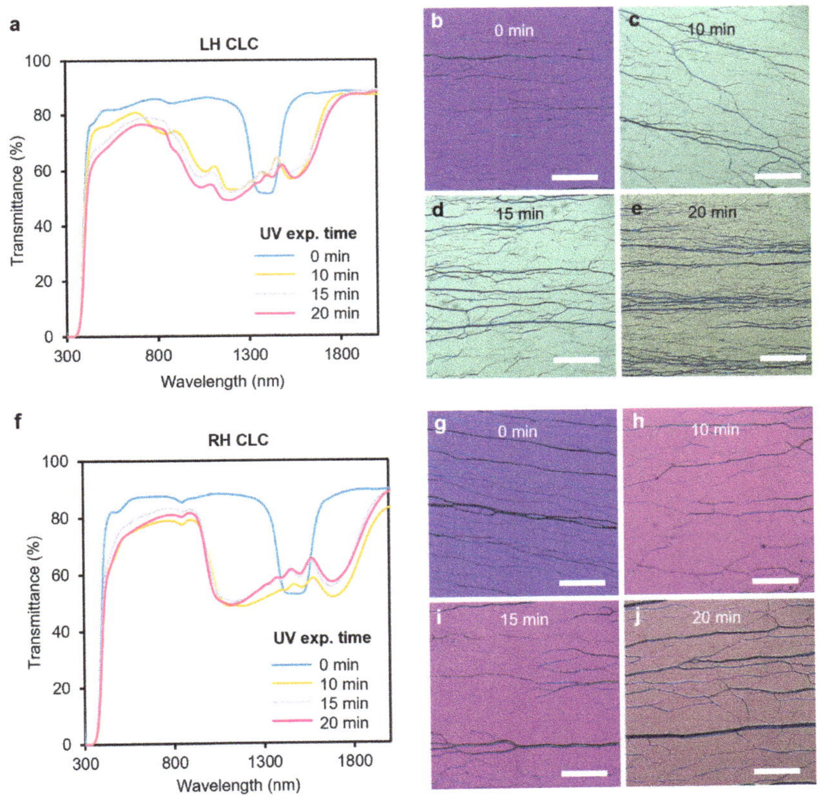

Figure 5. (**a**) Transmittance spectra of LH CLC layers under UV exposure times ranging from 0 to 20 min. (**b–e**) POM images of LH CLC layers with UV exposure times of (**b**) 0 min, (**c**) 10 min, (**d**) 15 min, and (**e**) 20 min. (**f**) Transmittance spectra of RH CLC layers under UV exposure times ranging from 0 to 20 min. (**g–j**) POM images of RH CLC layers with UV exposure times of (**g**) 0 min, (**h**) 10 min, (**i**) 15 min, and (**j**) 20 min. Scale bar, 500 µm.

A similar trend was observed for the RH CLC layer, where the transmittance spectra also saturated after 10 min of UV exposure (Figure 5f). POM images reveal a significant color change between 0 and 10 min of exposure (Figure 5g,h), stability between 10 and

15 min (Figure 5i), and a distinct color change after 20 min (Figure 5j). We selected 15 min as the optimized processing time to ensure complete polymerization and minimized potential instabilities caused by residual unreacted materials.

3.4. NIR Reflection Performance of the Stacked CLC Layers

To fabricate the final NIR-reflective film, the optimized LH and RH CLC layers were stacked together, as illustrated in Figure 2d. The stacked configuration exhibited high transmittance in the visible range (450–850 nm) and broadband reflection exceeding 50% in the NIR range (1000–1800 nm). The wavelength range between 850 and 1000 nm represents a transition from transmission to reflection. Upon lamination, the primary reflection peak of the individual CLC layers, initially centered near 1100 nm, shifted toward longer wavelengths. This redshift may be attributed to the diffusion of unreacted chiral dopants across the layer interface. The LH chiral dopant S811 remained unreacted, whereas the RH chiral dopant LC756 was fully consumed during polymerization. The resulting concentration gradient of S811 drove its migration into the RH CLC layer, partially relaxing the RH helical twisting and thereby shifting the reflection band with a new peak near 1200–1300 nm.

Durability to thermal stress was evaluated by heating the film up to 40 °C and 100 °C for 10 min. The film retained the visible light transmittance and NIR reflection properties after thermal treatment (Figure 6a). This excellent stability is primarily attributed to the acrylate-based thermoset polymer network formed by RM257. Although the host LC becomes optically isotropic at high temperatures, it preserves the LC phase due to stabilization within the polymer network. Removing E7 would degrade visible transmittance by increasing haze. The consistent transmittance spectra indicate that no evaporation or phase change of the host LC occurred upon heating.

To evaluate the durability to sunlight exposure, the NIR-reflective film was exposed to a broadband NIR light source (OSL2IR furnished with a cut-on filter at 550 nm, Thorlabs) ranging from 550 to 2000 nm (~30 mW/cm^2 at 980 nm) for durations between 4 and 48 h. As shown in Figure 6b, the transmittance spectra remained stable over the entire 48-h period, indicating no degradation in optical performance. Furthermore, the film exhibited no noticeable spectral degradation after being exposed to ambient conditions for over 500 h (Figure 6c). Considering the intrinsic hydrophobic nature of the constituent materials in the CLC layers, the film is expected to preserve its optical performance and potentially serve as an encapsulation layer [65,66].

To demonstrate the NIR-reflection performance of the stacked CLC layer, infrared thermography images of bare glass (Figure 6d) and the stacked CLC layer (Figure 6e) were captured under NIR illumination from above. The maximum temperatures recorded were 27.2 °C for bare glass and 43.4 °C for the stacked CLC layer. Since the recorded temperature correlates with the intensity of reflected NIR light, the stacked CLC layer exhibited higher NIR reflectance. In transmission mode, with NIR light incident from behind, a stronger NIR signal was detected passing through the bare glass (Figure 6f), whereas the stacked CLC layer effectively blocked the incoming NIR light (Figure 6g). The maximum temperatures recorded in this mode were 64.3 °C for bare glass and 34.7 °C for the stacked CLC layer.

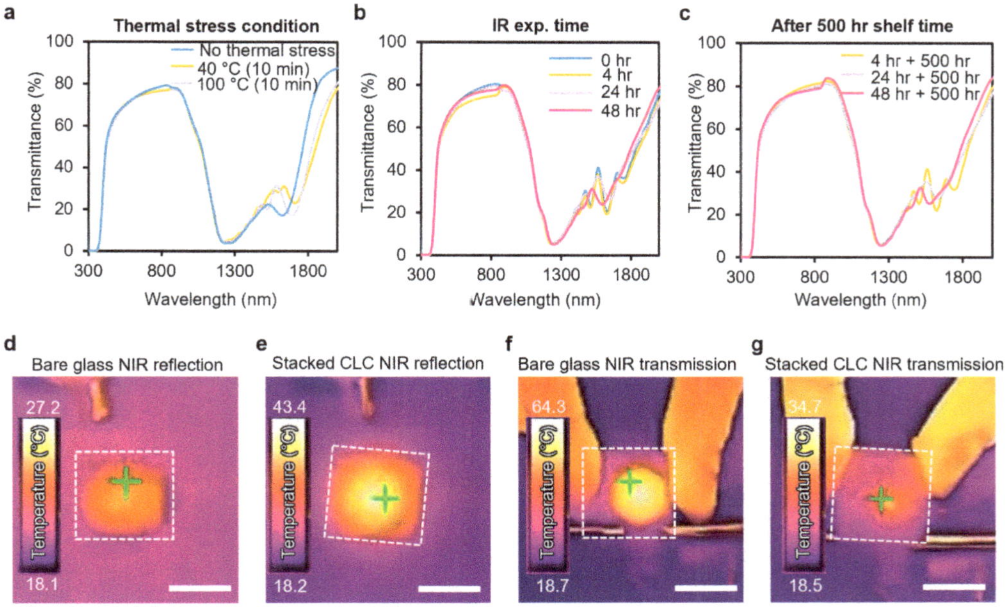

Figure 6. (**a**) Transmittance spectra of the stacked CLC layer before and after applying thermal stress at 40 °C and 100 °C for 10 min. (**b**) Transmittance spectra of the stacked CLC layer after NIR exposure for 0 to 48 h. (**c**) Transmittance spectra of the samples in (**b**) after being stored under ambient conditions for 500 h, following NIR exposure durations of 4, 24, and 48 h. (**d**,**e**) Infrared thermography images showing NIR reflection of (**c**) bare glass and (**d**) the stacked CLC layer (**d**). (**f**,**g**) Infrared thermography images showing NIR transmission through (**f**) bare glass and (**g**) the stacked CLC layer. In (**d**–**g**), a broadband NIR light source (550–2000 nm) with an intensity of 30 mW/cm^2 at 980 nm was used. Camera exposure time was automatically set between 8 and 12 ms. White dashed boxes indicate the sample positions. Scale bar, 2 cm.

4. Conclusions

We have developed a broadband NIR-reflective film composed of stacked LH and RH CLC layers. Through optimization of polymerizable monomer RM257 concentration, UV absorber UV-327 concentration, and UV exposure time, we realized a gradient-pitch helical structure with uniform phase in both LH and RH CLC layers. The optimized LH CLC layer was prepared with E7, RM257, S811, and UV-327 in weight ratios of 58.5:30:10:1, respectively (sample A4). The optimized RH CLC film was prepared with those materials in weight ratios of 66.5:30:2:1, respectively (sample B4). We selected the optimized UV intensity and exposure time to achieve a stable pitch gradient while preventing transmittance degradation caused by either incomplete polymerization or overexposure. The stacked CLC configuration achieved visible light transmission of up to 80% and broadband NIR reflection spanning 1000 to 1800 nm, with a maximum reflectance of 95%. The film demonstrated a high degree of durability under continuous sunlight exposure and thermal stress. This strategy of stacking two opposite-handed CLC layers offers an effective and versatile NIR-reflective solution suitable for energy-efficient and protective coatings in optoelectronic applications.

Author Contributions: Conceptualization, H.S.H., J.L., B.K. and S.-U.K.; methodology, J.L.; software, H.S.H. and J.L.; validation, B.K.; investigation, H.S.H., M.K. and D.K.; data curation, H.S.H., M.K. and D.K.; writing—original draft, H.S.H. and J.L.; writing—review and editing, J.L. and S.-U.K.;

visualization, J.L. and B.K.; supervision, S.-U.K.; project administration, S.-U.K.; funding acquisition, S.-U.K., H.S.H. and J.L. contributed equally to this study. All authors have read and agreed to the published version of the manuscript.

Funding: This study was financially supported by Seoul National University of Science and Technology.

Data Availability Statement: The raw data supporting the conclusions of this article will be made available by the authors on request.

Conflicts of Interest: The authors declare no conflicts of interest.

References

1. Wu, M.; Shi, Y.; Li, R.; Wang, P. Spectrally Selective Smart Window with High Near-Infrared Light Shielding and Controllable Visible Light Transmittance. *ACS Appl. Mater. Interfaces* **2018**, *10*, 39819–39827. [CrossRef]
2. Dang, S.; Wang, X.; Ye, H. An Ultrathin Transparent Radiative Cooling Photonic Structure with a High NIR Reflection. *Adv. Mater. Interfaces* **2022**, *9*, 2201050. [CrossRef]
3. Li, W.; Lin, C.; Ma, W.; Li, Y.; Chu, F.; Huang, B.; Yao, S. Transparent Selective Photothermal Coatings for Antifogging Applications. *Cell Rep. Phys. Sci.* **2021**, *2*, 100435. [CrossRef]
4. Herguedas, N.; Carretero, E. Evaluation of Low-Emissivity Coatings with Single, Double, and Triple Silver Layers. *Sol. Energy Mater. Sol. Cells* **2023**, *263*, 112592. [CrossRef]
5. Addonizio, M.L.; Ferrara, M.; Castaldo, A.; Antonaia, A. Air-Stable Low-Emissive AlN-Ag Based Coatings for Energy-Efficient Retrofitting of Existing Windows. *Energy Build.* **2021**, *250*, 111259. [CrossRef]
6. Capuzzo, G.; Kysylychyn, D.; Adhikari, R.; Li, T.; Faina, B.; Tarazaga Martín-Luengo, A.; Bonanni, A. All-Nitride AlxGa1−xN:Mn/GaN Distributed Bragg Reflectors for the near-Infrared. *Sci. Rep.* **2017**, *7*, 42697. [CrossRef]
7. Yan, X.; Wang, J.; Zhang, W.; Liu, Y.; Luo, D. Gradient Polarization Volume Grating with Wide Angular Bandwidth for Augmented Reality. *Opt. Express* **2023**, *31*, 35282. [CrossRef]
8. Dalapati, G.K.; Masudy-Panah, S.; Chua, S.T.; Sharma, M.; Wong, T.I.; Tan, H.R.; Chi, D. Color Tunable Low Cost Transparent Heat Reflector Using Copper and Titanium Oxide for Energy Saving Application. *Sci. Rep.* **2016**, *6*, 20182. [CrossRef]
9. Yu, H.; Xu, G.; Shen, X.; Yan, X.; Cheng, C. Low Infrared Emissivity of Polyurethane/Cu Composite Coatings. *Appl. Surf. Sci.* **2009**, *255*, 6077–6081. [CrossRef]
10. Wurm, J.; Fujisawa-Phillips, S.T.; Rasskazov, I.L. Optimal Design of Low-Emissivity Coatings. *Sol. Energy Mater. Sol. Cells* **2025**, *280*, 113267. [CrossRef]
11. Gryga, M.; Ciprian, D.; Hlubina, P. Distributed Bragg Reflectors Employed in Sensors and Filters Based on Cavity-Mode Spectral-Domain Resonances. *Sensors* **2022**, *22*, 3627. [CrossRef]
12. Mansour, S.A.; Farha, A.H. A Review of Near-Infrared Reflective Nanopigments: Aesthetic and Cooling Properties. *Crystals* **2025**, *15*, 271. [CrossRef]
13. Ahmed, U.; Khan, Y.; Ehsan, M.; Amirzada, M.; Ullah, N.; Khatri, A.; Ur Rehman, A.; Butt, M. Investigation of Spectral Properties of DBR-Based Photonic Crystal Structure for Optical Filter Application. *Crystals* **2022**, *12*, 409. [CrossRef]
14. Peng, Y.; Zhang, L.; Cheng, N.; Andrew, T. ITO-Free Transparent Organic Solar Cell with Distributed Bragg Reflector for Solar Harvesting Windows. *Energies* **2017**, *10*, 707. [CrossRef]
15. Ahn, S.-C.; Lee, B.-T.; An, W.-C.; Kim, D.-K.; Jang, I.-K.; So, J.-S.; Lee, H.-J. Optimum Conditions of the Distributed Bragg Reflector in 850-Nm GaAs Infrared Light-Emitting Diodes. *J. Korean Phys. Soc.* **2016**, *69*, 91–95. [CrossRef]
16. Liu, L.; Chang, H.; Xu, T.; Song, Y.; Zhang, C.; Hang, Z.H.; Hu, X. Achieving Low-Emissivity Materials with High Transmission for Broadband Radio-Frequency Signals. *Sci. Rep.* **2017**, *7*, 4840. [CrossRef]
17. Shaaban, I.E.; Samra, A.S.; Muhammad, S.; Wageh, S. Design of Distributed Bragg Reflectors for Green Light-Emitting Devices Based on Quantum Dots as Emission Layer. *Energies* **2022**, *15*, 1237. [CrossRef]
18. Wu, C.M.; Zhang, B.P.; Shang, J.Z.; Cai, L.E.; Zhang, J.Y.; Yu, J.Z.; Wang, Q.M. High-Reflectivity AlN/GaN Distributed Bragg Reflectors Grown on Sapphire Substrates by MOCVD. *Semicond. Sci. Technol.* **2011**, *26*, 055013. [CrossRef]
19. Gao, B.; George, J.P.; Beeckman, J.; Neyts, K. Design, Fabrication and Characterization of a Distributed Bragg Reflector for Reducing the Étendue of a Wavelength Converting System. *Opt. Express* **2020**, *28*, 12837. [CrossRef]
20. Kim, H.; Kaya, M.; Hajimirza, S. Broadband Solar Distributed Bragg Reflector Design Using Numerical Optimization. *Sol. Energy* **2021**, *221*, 384–392. [CrossRef]
21. Prateek; Huang, T.; Uchiyama, T.; Jung, D.E.; Verrastro, L.D.; Yan, J.; Brandso, S.; Stewart, B.; Katsumata, R.; Nuguri, S.M.; et al. High-Efficiency Bragg Mirrors by Solution-Based and Roll-to-Roll Processing. *Nano Lett.* **2025**, *25*, 4713–4719. [CrossRef]
22. Zhan, X.; Xu, F.; Zhou, Z.; Yan, Y.; Yao, J.; Zhao, Y.S. 3D Laser Displays Based on Circularly Polarized Lasing from Cholesteric Liquid Crystal Arrays. *Adv. Mater.* **2021**, *33*, 2104418. [CrossRef]

23. Dinc, R.U.; Lub, J.; Kragt, A.J.J.; Schenning, A.P.H.J. An L-Isosorbide-Based Reactive Chiral Dopant with High Helical Twisting Power for Cholesteric Liquid Crystal Polymers Reflecting Left-Handed Circularly Polarized Light. *Org. Chem. Front.* **2024**, *11*, 7053–7058. [CrossRef]
24. Li, H.; Xu, J.; Ren, Y.; Han, R.; Song, H.; Huang, R.; Wang, X.; Zhang, L.; Cao, H.; Zou, C.; et al. Preparation of Highly Durable Reverse-Mode Polymer-Stabilized Liquid Crystal Films with Polymer Walls. *ACS Appl. Mater. Interfaces* **2023**, *15*, 2228–2236. [CrossRef]
25. Lee, K.M.; Tondiglia, V.P.; McConney, M.E.; Natarajan, L.V.; Bunning, T.J.; White, T.J. Color-Tunable Mirrors Based on Electrically Regulated Bandwidth Broadening in Polymer-Stabilized Cholesteric Liquid Crystals. *ACS Photonics* **2014**, *1*, 1033–1041. [CrossRef]
26. Broer, D.J.; Mol, G.N.; van Haaren, J.A.M.M.; Lub, J. Photo-Induced Diffusion in Polymerizing Chiral-Nematic Media. *Adv. Mater.* **1999**, *11*, 573–578. [CrossRef]
27. Tamaoki, N.; Moriyama, M.; Matsuda, H. Cholesteric Solid Films Formed by Spin-Coating Solutions of Dicholesteryl Esters. *Angew. Chem. Int. Ed.* **2000**, *39*, 509–511. [CrossRef]
28. Maeng, J.; Rihani, R.T.; Javed, M.; Rajput, J.S.; Kim, H.; Bouton, I.G.; Criss, T.A.; Pancrazio, J.J.; Black, B.J.; Ware, T.H. Liquid Crystal Elastomers as Substrates for 3D, Robust, Implantable Electronics. *J. Mater. Chem. B* **2020**, *8*, 6286–6295. [CrossRef]
29. Mulder, D.J.; Schenning, A.P.H.J.; Bastiaansen, C.W.M. Chiral-Nematic Liquid Crystals as One Dimensional Photonic Materials in Optical Sensors. *J. Mater. Chem. C* **2014**, *2*, 6695–6705. [CrossRef]
30. Li, S.-L.; Chen, Z.-Y.; Chen, P.; Hu, W.; Huang, C.; Li, S.-S.; Hu, X.; Lu, Y.-Q.; Chen, L.-J. Geometric Phase-Encoded Stimuli-Responsive Cholesteric Liquid Crystals for Visualizing Real-Time Remote Monitoring: Humidity Sensing as a Proof of Concept. *Light Sci. Appl.* **2024**, *13*, 27. [CrossRef]
31. Li, X.; Chen, Y.; Du, C.; Liao, X.; Yang, Y.; Feng, W. Cholesteric Liquid Crystal Elastomer Coatings with Brilliant Structural Colors and Mechanochromic Response Fabricated by Spray Deposition. *Adv. Funct. Mater.* **2025**, *35*, 2412298. [CrossRef]
32. Picot, O.T.; Dai, M.; Broer, D.J.; Peijs, T.; Bastiaansen, C.W.M. New Approach toward Reflective Films and Fibers Using Cholesteric Liquid-Crystal Coatings. *ACS Appl. Mater. Interfaces* **2013**, *5*, 7117–7121. [CrossRef] [PubMed]
33. Kim, D.-Y.; Lee, K.M.; White, T.J.; Jeong, K.-U. Cholesteric Liquid Crystal Paints: In Situ Photopolymerization of Helicoidally Stacked Multilayer Nanostructures for Flexible Broadband Mirrors. *NPG Asia Mater.* **2018**, *10*, 1061–1068. [CrossRef]
34. Zhang, W.; Lub, J.; Schenning, A.P.H.J.; Zhou, G.; de Haan, L.T. Polymer Stabilized Cholesteric Liquid Crystal Siloxane for Temperature-Responsive Photonic Coatings. *Int. J. Mol. Sci.* **2020**, *21*, 1803. [CrossRef]
35. Tondiglia, V.P.; Natarajan, L.V.; Bailey, C.A.; McConney, M.E.; Lee, K.M.; Bunning, T.J.; Zola, R.; Nemati, H.; Yang, D.-K.; White, T.J. Bandwidth Broadening Induced by Ionic Interactions in Polymer Stabilized Cholesteric Liquid Crystals. *Opt. Mater. Express* **2014**, *4*, 1465. [CrossRef]
36. Belalia, M.; Mitov, M.; Bourgerette, C.; Krallafa, A.; Belhakem, M.; Bormann, D. Cholesteric Liquid Crystals with a Helical Pitch Gradient: Spatial Distribution of the Concentration of Chiral Groups by Raman Mapping in Relation with the Optical Response and the Microstructure. *Phys. Rev. E* **2006**, *74*, 051704. [CrossRef]
37. Zografopoulos, D.C.; Kriezis, E.E.; Mitov, M.; Binet, C. Theoretical and Experimental Optical Studies of Cholesteric Liquid Crystal Films with Thermally Induced Pitch Gradients. *Phys. Rev. E* **2006**, *73*, 061701. [CrossRef]
38. Li, F.; Wang, L.; Sun, W.; Liu, H.; Liu, X.; Liu, Y.; Yang, H. Dye Induced Great Enhancement of Broadband Reflection from Polymer Stabilized Cholesteric Liquid Crystals. *Polym. Adv. Technol.* **2012**, *23*, 143–148. [CrossRef]
39. Shi, W.; Zhang, X.; Han, R.; Li, H.; Cao, H.; Chen, Y.; Wang, D.; Yang, Z.; He, W. Preparation of Cholesteric Polymer Networks with Broadband Reflection Memory Effect. *Liq. Cryst.* **2022**, *49*, 153–161. [CrossRef]
40. Zhang, H.; Yu, P.; Zhong, T.; Zhao, Y.; Miao, Z.; He, Z.; Li, K.; Zhang, Y.; Zhao, Y.; Shen, W. Preparation of Chiral Polymer/Cholesteric Liquid Crystals Composite Films with Broadband Reflective Capability for Smart Windows and Thermal Management of Buildings. *Opt. Mater.* **2021**, *121*, 111611. [CrossRef]
41. Zhou, H.; Hu, X.; Liu, M.; Yin, D. Benzotriazole Ultraviolet Stabilizers in the Environment: A Review of Analytical Methods, Occurrence, and Human Health Impacts. *TrAC Trends Anal. Chem.* **2023**, *166*, 117170. [CrossRef]
42. Fischer, C.; Leibold, E.; Göen, T. Identification of in Vitro Phase I Metabolites of Benzotriazole UV Stabilizer UV-327 Using HPLC Coupled with Mass Spectrometry. *Toxicol. In Vitro* **2020**, *68*, 104932. [CrossRef] [PubMed]
43. Song, J.; McCormick, A.V.; Francis, L.F. Depthwise Viscosity Gradients in UV-cured Epoxy Coatings. *Macromol. Mater. Eng.* **2013**, *298*, 145–152. [CrossRef]
44. Liu, Y.; Yuan, X.; Xie, M.; Liu, Z.; Zhao, X.; Cao, H.; Wang, H.; Yang, Z.; Wang, D.; He, W. Flexible, Easy-to-Produce, Gradient Distributed Pitch Broadband Infrared Reflectors with Polymer-Stabilized Cholesteric Liquid Crystals. *J. Mol. Liq.* **2024**, *408*, 125369. [CrossRef]
45. Zhang, L.; He, W.; Yuan, X.; Hu, W.; Cao, H.; Yang, H.; Zhu, S. Broadband Reflection Characteristic of Polymer-Stabilised Cholesteric Liquid Crystal with Pitch Gradient Induced by a Hydrogen Bond. *Liq. Cryst.* **2010**, *37*, 1275–1280. [CrossRef]

46. Rojas-Rodríguez, M.; Fiaschi, T.; Mannelli, M.; Mortati, L.; Celegato, F.; Wiersma, D.S.; Parmeggiani, C.; Martella, D. Cellular Contact Guidance on Liquid Crystalline Networks with Anisotropic Roughness. *ACS Appl. Mater. Interfaces* **2023**, *15*, 14122–14130. [CrossRef]
47. Zhang, J.; Tian, Z.; Ji, X.; Zhang, F. Fabrication Mechanisms of Lignin Nanoparticles and Their Ultraviolet Protection Ability in PVA Composite Film. *Polymers* **2022**, *14*, 4196. [CrossRef]
48. Hu, X.; Zeng, W.; Zhang, X.; Wang, K.; Liao, X.; Jiang, X.; Jiang, X.-F.; Jin, M.; Shui, L.; Zhou, G. Pitch Gradation by Ion-Dragging Effect in Polymer-Stabilized Cholesteric Liquid Crystal Reflector Device. *Polymers* **2020**, *12*, 96. [CrossRef]
49. Choi, J.-H.; Hwang, H.S.; Jang, H.-B.; Kim, S.-U.; Park, H.-L. Flexible Phototransistors Integrated with Chiral Liquid Crystal Encapsulating Film for Improving Color Selectivity and Stability. *ACS Appl. Electron. Mater.* **2024**, *6*, 8094–8103. [CrossRef]
50. Park, H.-L.; Jun, J.; Kim, M.-H.; Lee, S.-H. Introduction of Helical Photonic Crystal Insulator in Organic Phototransistor for Enhancing Selective Color Absorption. *Org. Electron.* **2022**, *100*, 106385. [CrossRef]
51. Yang, C.-S.; Lin, C.-J.; Pan, R.-P.; Que, C.T.; Yamamoto, K.; Tani, M.; Pan, C.-L. The Complex Refractive Indices of the Liquid Crystal Mixture E7 in the Terahertz Frequency Range. *J. Opt. Soc. Am. B* **2010**, *27*, 1866. [CrossRef]
52. Sathaye, K.S. Asymmetric Tunable Fabry-Perot Cavity Using Switchable Polymer Stabilized Cholesteric Liquid Crystal Optical Bragg Mirror. *Opt. Eng.* **2012**, *51*, 034001. [CrossRef]
53. Phillips, A.T.; Chen, J.C.; McCracken, J.M.; White, T.J. Dynamic Infrared Reflective Filters Prepared from Cholesteric Liquid Crystalline Elastomers. *ACS Appl. Opt. Mater.* **2024**, *2*, 2559–2567. [CrossRef]
54. Sol, J.A.H.P.; Timmermans, G.H.; van Breugel, A.J.; Schenning, A.P.H.J.; Debije, M.G. Multistate Luminescent Solar Concentrator "Smart" Windows. *Adv. Energy Mater.* **2018**, *8*, 1702922. [CrossRef]
55. Duda, Ł.; Czajkowski, M.; Potaniec, B.; Vaňkátová, P. Helical Twisting Power and Compatibility in Twisted Nematic Phase of New Chiral Liquid Crystalline Dopants with Various Liquid Crystalline Matrices. *Liq. Cryst.* **2019**, *46*, 1769–1779. [CrossRef]
56. Yoshida, J.; Tamura, S.; Hoshino, K.; Yuge, H.; Sato, H.; Yamazaki, A.; Yoneda, S.; Watanabe, G. Comprehensive Understanding of Host- and Guest-Dependent Helix Inversion in Chiral Nematic Liquid Crystals: Experimental and Molecular Dynamics Simulation Study. *J. Phys. Chem. B* **2018**, *122*, 10615–10626. [CrossRef] [PubMed]
57. Zhan, X.; Fan, H.; Li, Y.; Liu, Y.; Luo, D. Low Threshold Polymerised Cholesteric Liquid Crystal Film Lasers with Red, Green and Blue Colour. *Liq. Cryst.* **2019**, *46*, 970–976. [CrossRef]
58. Lub, J.; Broer, D.J.; Wegh, R.T.; Peeters, E.; Van Der Zande, B.M.I. Formation of Optical Films by Photo-Polymerisation of Liquid Crystalline Acrylates and Application of These Films in Liquid Crystal Display Technology. *Mol. Cryst. Liq. Cryst.* **2005**, *429*, 77–99. [CrossRef]
59. Agez, G.; Mitov, M. Cholesteric Liquid Crystalline Materials with a Dual Circularly Polarized Light Reflection Band Fixed at Room Temperature. *J. Phys. Chem. B* **2011**, *115*, 6421–6426. [CrossRef]
60. Choi, T.-H.; Do, S.-M.; Jeon, B.-G.; Shin, S.T.; Yoon, T.-H. Formation of Polymer Walls through the Phase Separation of a Liquid Crystal Mixture Induced by a Spatial Elastic Energy Difference. *Sci. Rep.* **2019**, *9*, 10288. [CrossRef]
61. Zhan, X.; Luo, D.; Yang, K.-L. Multifunctional Sensors Based on Liquid Crystals Scaffolded in Nematic Polymer Networks. *RSC Adv.* **2021**, *11*, 38694–38702. [CrossRef] [PubMed]
62. Guo, R.; Li, K.; Cao, H.; Wu, X.; Wang, G.; Cheng, Z.; Wang, F.; Zhang, H.; Yang, H. Chiral Polymer Networks with a Broad Reflection Band Achieved with Varying Temperature. *Polymer* **2010**, *51*, 5990–5996. [CrossRef]
63. Lin, H.; Zhao, Y.; Gao, H.; Du, Z.; Zhang, H.; Luan, Y.; Wang, D.; Li, C. Preparation and Application of Broadband Reflective Polymer-Stabilised Cholesteric Liquid with Nanoscale Fibres. *Liq. Cryst.* **2023**, *50*, 2280–2294. [CrossRef]
64. Zhang, X.; Shi, W.; Han, R.; Li, H.; Cao, H.; Chen, Y.; Yang, Z.; Wang, D.; He, W. Self-Diffusion Method for Broadband Reflection in Polymer-Stabilized Cholesteric Liquid Crystal Films. *Liq. Cryst.* **2022**, *49*, 494–503. [CrossRef]
65. Xia, X.; Peng, J.; Wan, Q.; Wang, X.; Fan, Z.; Zhao, J.; Li, F. Functionalized Ionic Liquid-Crystal Additive for Perovskite Solar Cells with High Efficiency and Excellent Moisture Stability. *ACS Appl. Mater. Interfaces* **2021**, *13*, 17677–17689. [CrossRef]
66. Hwang, G.-T.; Im, D.; Lee, S.E.; Lee, J.; Koo, M.; Park, S.Y.; Kim, S.; Yang, K.; Kim, S.J.; Lee, K.; et al. In Vivo Silicon-Based Flexible Radio Frequency Integrated Circuits Monolithically Encapsulated with Biocompatible Liquid Crystal Polymers. *ACS Nano* **2013**, *7*, 4545–4553. [CrossRef]

crystals

MDPI

Article

Maintaining Glycerol-Based Hexagonal Structures by Crosslinkers for High Permeability Nanofiltration

Senlin Gu [1,2,3], Luke A. O'Dell [3] and Lingxue Kong [3,*]

1 Hangzhou Institute for Advanced Study, University of Chinese Academy of Sciences, Hangzhou 310024, China; gusenlin@hotmail.com
2 School of Engineering, Westlake University, Hangzhou 310024, China
3 Institute for Frontier Materials, Deakin University, Geelong, VIC 3216, Australia; luke.odell@deakin.edu.au
* Correspondence: lingxue.kong@deakin.edu.au

Abstract

Hypothesis: Structural optimization of crosslinkers within a reactive glycerol-based hexagonal lyotropic liquid crystal (HLLC) system is proposed to enhance the interfacial stability of hexagonal mesophases and improve the hexagonal structure retention during polymerization. This targeted modification is anticipated to significantly improve the water filtration efficiency of HLLC-templated nanofiltration. **Experiments:** The effect of crosslinkers on the interfacial stability of glycerol-based hexagonal mesophases was studied by evaluating their concentration accommodation within the mesophases using ^{13}C solid NMR, FTIR and SAXS. **Findings:** A hydrophilic crosslinker consisting of ten ethylene glycol units shows less interference with the interfacial stability of hexagonal mesophases, therefore contributing to a higher concentration accommodation compared to the one with three ethylene glycol units. This long-chain crosslinker, despite having a low content of reactive groups, effectively connects the cylinders and better retains the hexagonal structures during polymerization than the hydrophobic crosslinker with shorter ethylene glycol units but a higher content of reactive groups. The retained hexagonal nanofiltration membranes show a remarkable pure water permeability of 40 L m^{-2} h^{-1} bar^{-1} μm, resulting from the strong hygroscopic effect of glycerol and the crumpled surface of membranes due to the flexible nature of the system plasticized by glycerol.

Keywords: nanofiltration; hexagonal lyotropic liquid crystal; interfacial stability; crosslinkers; glycerol

1. Introduction

Nanofiltration membranes synthesized through hexagonal lyotropic liquid crystals (HLLC) hold promise for bolstering membrane efficiency through their unique blend of superior selectivity and permeability [1–5]. Despite their ability to generate tunable nanopore geometries and high pore densities, lyotropic liquid crystals (LLC) face significant stability challenges in aqueous mesophases, limiting their implementation in water filtration membranes. Studies indicate that incorporating polar organic solvents can stabilize LLC mesophases in diverse surfactant and lipid systems [6–9]. Compared to other solvents, glycerol, a nontoxic water miscible solvent with low volatility (normal boiling point 290 °C; vapor pressure @ 20 °C < 1 Torr) [1] has been used as pure solvent or with water to engineer higher-stability LLC mesophases. However, because of the lower polarity of glycerol than water, mesophases, such as hexagonal, cubic or lamellar ones, usually need to be formed in

https://doi.org/10.3390/cryst15070664

temperature ranges much higher than room temperature [10–14]. This could increase the difficulties for later membrane fabrication, since many polymeric porous substrates used for supporting the LLC-templated active layer are sensitive to temperature.

Glycerol-based LLC systems have been developed at room temperature and showed better performance compared to water-based ones [15,16]. Nonionic ethoxylated phytosterol (BPS) surfactants with various oxyethylene unit lengths were reported to form lamellar mesophase with glycerol at room temperature. It was found that the BPS/glycerol systems possess wider concentration ranges of LLC mesophases and higher viscosity than BPS/water systems [15], which is because the H-bonding mean lifetime of glycerol is longer than that of water [17,18]. Shukla et al. reported the formation of hexagonal and lamellar phase by sodium dodecyl sulphate (SDS) and glycerol at room temperature [16]. Compared to water based-LLC mesophases, glycerol-based LLC mesophases are more ordered and compact in the literature due to the stronger solvophobic effect caused by the longer H-bonding mean lifetime [19].

Low-volatility glycerol-based LLC systems are promising for fabricating (solution-based processing) nanofiltration membranes with good LLC-templated structure retention [6,7]. Nanofiltration membranes with cubic mesophase (Q_I)-templated structures were fabricated at 70 °C using an alkydiene-imidazolium bromide-based gemini surfactant/glycerol system. This Q_I membrane shows a remarkable $MgCl_2$ rejection of 99% but a low thickness-normalized permeability of 0.066 L m^{-2} h^{-1} bar^{-1} μm [9]. Osuji et al. fabricated glycerol-based HLLC-templated membranes formed by mono-acrylate quaternary surfactants at room temperature. These membranes showcase tunable pore sizes in a range of 0.6–1.5 nm and a water permeability higher than 2 L m^{-2} h^{-1} bar^{-1} μm [1,7]. The synthesis of mono-acrylate quaternary surfactants involves only one step [2,7]. These kinds of surfmers were reported to form hexagonal mesophases with glycerol easily at room temperature [1,7], and the hexagonal structures can be retained with the assistance of crosslinkers [2].

The extensive chemical diversity of commercially available crosslinkers yields a substantial combinatorial space of potential molecular candidates [20,21]. However, molecular-level optimization strategies for glycerol-based HLLC template retention remain unexplored. Insight into the structural effect of crosslinkers on the interfacial stability of glycerol-based hexagonal structures before polymerization and the retention of hexagonal structures during polymerization is crucial to optimize the HLLC-based nanofiltration membranes.

Herein, the HLLC system of polymerizable surfactant 2-(acryloyloxy) ethyl dodecyl dimethyl ammonium bromide (AED) (Figure 1) and glycerol acts as the binary template [1]. It was found that the AED/glycerol system possesses a broad range of hexagonal phases. The pure hexagonal binary system with the highest concentration of reactive groups was filtered. Crosslinkers were strategically incorporated into the binary system to modulate macroscopic phase behavior and stabilize the template architecture during polymerization. Selection criteria prioritized (i) hydrophilic character, governing crosslinker localization within the mesophase, and (ii) molecular length, directly controlling crosslinking density. The subtle variations in molecular structures are responsible for the interfacial stability of the formed ternary hexagonal mesophases. HLLC structure retention was achieved by introducing a specific hydrophilic crosslinker, which mediated the structure retention mechanism. The variations in the ordering and unit cell sizes of HLLC structures are responsible for the transparent and rejection properties. The variations in the surface structures of HLLC-templated membranes are responsible for high water permeability.

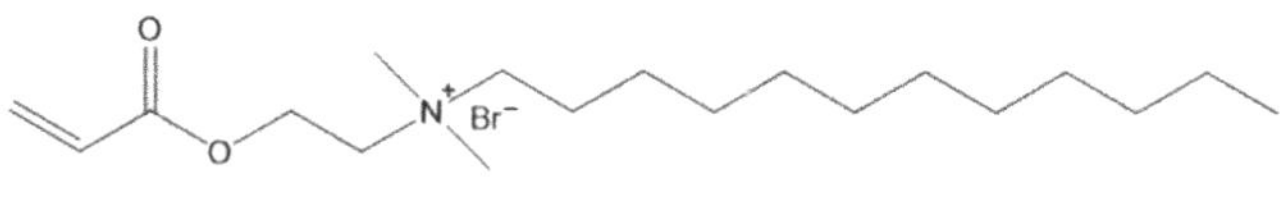

2-(acryloyloxy) ethyl dodecyl dimethyl ammonium bromide (AED)

Figure 1. Chemical structure of polymerizable surfactant AED.

2. Materials and Methods

2.1. Materials

The following reagents were sourced from Sigma-Aldrich and used as received (inhibitor was removed before using): (i) *Crosslinkers*: Poly (ethylene glycol) diacrylate (PEGDA250, M_n = 250 g/mol; PEGDA575, M_n = 575 g/mol), pentaerythritol tetraacrylate (PETA); (ii) *Photoinitiator*: 2-Hydroxy-2-methylpropiophenone (HMPP); (iii) *Surfactant precursors*: 2-(Dimethylamino)ethyl acrylate and 1-bromododecane (subsequently copolymerized to form quaternary ammonium surfactant AED). The quaternary ammonium surfactant AED was prepared using modified literature procedures, with full synthetic characterization provided in Section S1 [1,7].

2.2. LLC Formation and Crosslinking

Homogeneous LLC precursor mixtures containing surfactant, crosslinker, glycerol and photoinitiator (0.5 wt% relative to total solids) were prepared in amber vials. Thermal processing at 70–80 °C facilitated mesophase formation, with intermittent vortex mixing and centrifugation accelerating homogenization. LLC sample compositions are shown in Table S1.

A 365 nm UV source was employed to cure samples [22]. Bulk HLLC samples were UV-cured (20 min) between fluorinated ethylene propylene (FEP) spacers (0.15 mm thick) positioned 1 cm from the UV source. For thin-film composite (TFC) membranes, HLLC precursor gel was mechanically cast onto polyacrylonitrile ultrafiltration (UPAN) substrates using polyester films (silicon oil coated) as spacers, followed by N_2-purging (10 min) and UV-curing (365 nm, 20 min) under continuous N_2 flow. The TFC membranes were marked as AED76@PAN and AED76P2@PAN (AED76 and AED76P2 active layers after polymerization on the UPAN substrate).

2.3. Characterizations

SAXS measurements were conducted by using a Xeuss 3.0 system (Xenocs, Grenoble, France) to identify the lyotropic liquid crystal (LLC) phase structure pre and post polymerization. Around 50 mg of sample was loaded in a SAXS cell and covered by a thin Kapton film. A dual-source configuration was utilized: a GeniX3D Ga source (Deakin-CSIRO, λ = 1.34 Å) and a Cu source (ANSTO, λ = 1.54 Å). Data were collected at a sample-to-detector distance of 350 mm, covering a scattering vector range (q) of ~0.1 to 0.75 Å^{-1}. Exposure times were set between 300 s and 1800 s. Two-dimensional scattering patterns were acquired in line-eraser mode and subsequently calibrated using silver behenate, then radially averaged to yield one-dimensional intensity profiles. Bragg peaks were subjected to Gaussian fitting analysis using XSACT (v2.7) software (Xenocs) to extract scattered intensity (I) as a function of q.

The cross-sectional morphology of the HLLC layer was observed using scanning electron microscopy (SEM; Zeiss Supra 55VP, Overkochen, Germany). Samples were cryofractured in liquid nitrogen and sputter-coated with platinum prior to SEM imaging.

Triplicate measurements were performed on three distinct specimens to ensure representative characterization.

The evolution of birefringent textures in the HLLC was characterized using polarized optical microscopy (POM; Nikon Eclipse 80i, Tokyo, Japan).

The tensile mechanical properties of polymerized LLC-templated materials were determined using an Instron 5967 universal testing system. Rectangular specimens (nominal dimensions: 18 mm × 4 mm) were tested at a crosshead speed of 5 mm/min, employing a 5 N load cell and mechanical grips.

^{13}C solid-state nuclear magnetic resonance (NMR) spectra were acquired on a Bruker Avance III spectrometer (11.7 T) (Billerica, MA, USA) using a static 5 mm probe. Measurements were performed without ^1H decoupling employing the following parameters: recycle delay = 1 s, pw90 = 7 μs (at 100 W power). Samples were loaded into 5 mm outer diameter NMR tubes (length: 3–4 cm) and sealed with laboratory film and caps prior to analysis.

UV-Vis transmittance spectra of thin HLLC films were recorded on an Agilent Cary 5000 (Santa Clara, CA, USA) spectrophotometer over 400–800 nm.

Polymerization of the HLLC membranes was characterized by Fourier transform infrared spectroscopy (FTIR) using a Bruker Vertex 70 (Billerica, MA, USA) spectrometer equipped with an ATR accessory. Spectra were acquired at 2 cm^{-1} resolution with 64 co-added scans across 4000–400 cm^{-1}.

Water permeability and ion rejection of TFC membranes were characterized using a crossflow nanofiltration system (Saifei SF-SA, Hangzhou, China) with an effective membrane area of 7.07 cm^2. Following standardized preconditioning (5.5 bar, 5 min), permeability measurements were conducted at 25 °C and 5.5 bar transmembrane pressure. Deionized water-wetted membranes were securely mounted in the test cell prior to all measurements. Water permeance (J) was calculated using Equation (1) [23]:

$$J = \frac{\Delta w}{\rho A \Delta t P} \quad (1)$$

where permeability parameters are defined as follows: Δw: mass of permeate collected (g); Δt: collection time interval (h); A: effective membrane area (m^2); ρ: permeate density (g/L); P: applied transmembrane pressure (bar).

Solute rejection (R) was computed using Equation (2) [23]:

$$R_1 = \left(1 - \frac{C_p}{C_f}\right) \times 100\% \quad (2)$$

where C_p and C_f represent permeate and feed conductivities, respectively.

3. Results and Discussion

3.1. The Effect of Crosslinkers on the Interfacial Stability of the HLLC Mesophases Before Polymerization

The structural optimization of crosslinkers can enhance the interfacial stability of hexagonal mesophases before polymerization, which determines the concentration accommodation of crosslinkers in HLLC systems. SAXS and polarized optical microscopy have been used to study the mesophase structure of AED/glycerol systems (Figure S3). The primary peak (q = ~0.2 $Å^{-1}$) corresponding to the HLLC phase is significant in most of the samples (from 55 to 94 wt% of AED), indicating that all of them possess hexagonal phase. The hexagonal phase formation for the same systems has been verified before [1]. However, the 1D SAXS patterns of the AED55 (AED/glycerol = 55/45, w/w), AED60 (AED/glycerol = 60/40, w/w) and AED65 (AED/glycerol = 65/35, w/w) samples show a two-phase system, where a broad isotropic peak appears with the primary peak of HLLC. Moreover, a two-phase system appears again for samples AED88 (AED/glycerol = 88/12, *w*/*w*) and AED94 (AED/glycerol = 94/6, *w*/*w*), where AED crystal peaks appear in 1D SAXS patterns. Therefore, AED71 (AED/glycerol = 71/29, *w*/*w*), AED76 (AED/glycerol = 76/24, *w*/*w*) and AED82 (AED/glycerol = 82/18, *w*/*w*) are pure hexagonal phases identified from SAXS patterns. Additionally, from POM, more birefringence textures corresponding to hexagonal phase appear from adding 55 to 66 wt%, indicating the increasing proportion of hexagonal phase in the system. Crystalline textures start to appear in the AED82 system, indicating the coexistence of hexagonal phase and crystal in this system. The AED crystal amount in AED82 may be so small that the corresponding Bragg peak could not be very significant in terms of SAXS pattern. Therefore, only AED71 and AED76 are single hexagonal phase systems.

Since higher concentrations of reactive groups in the HLLC precursor can better retain the HLLC structure during polymerization, the AED76 binary system has been used to form a ternary system with crosslinkers. Three crosslinkers (Figure 2A) with various molecular lengths and hydrophilicities were introduced into the AED76 system individually. The introduction of crosslinker exerts great effects on the stability of HLLC phase as shown in Figures 2 and 3. It is found that with the increasing concentration of crosslinker, the intensity of primary peaks becomes lower and broad isotropic peaks become more significant, indicating the gradual destabilization of mesophase interfaces and the transition from a single hexagonal phase to a two-phase system. For the AED76 systems with PETA or PEGDA575, the intensity of hexagonal primary peaks decreases significantly and the broad isotropic peaks appear when adding more than 2 wt% of crosslinkers, while for the AED76 system with PEGDA250, this phenomenon occurs when adding 2 wt% of PEGDA250. Therefore, PEGDA575 and PETA can better stabilize the interfaces than PEGDA250 and the AED76 system with stable hexagonal mesophase formation can contain 2 wt% of PEGDA575 or PETA. For the AED76P2 and AED76PE2 systems, the lattice parameter has no change after adding PEGDA or PETA into the system, as shown in Table 1. However, the radius of lipid core R decreases and the interfacial area per hydrophilic head group of the surfactant S increases after incorporating crosslinkers, especially for the system with PETA. These variations indicate the increasing interfacial curvature of cylinders after the addition of crosslinkers, thereby resulting in a looser and more unstable packing of the surfactant molecules. The much smaller R and larger S of sample AED76PE2 (AED/glycerol/PETA = 76/22/2, *w*/*w*) compared to AED76P2 (AED/glycerol/PEGDA575 = 76/22/2, *w*/*w*) suggests that low-polarity PETA exerts greater effects on the packing of surfactant molecules.

From the POM images shown in Figure 2, uniform birefringent textures can be found in the AED76PE2 and AED76P2 systems, while some areas become non-birefringent after adding more than 2 wt% of PEGDA575 or PETA into the AED76 system. These results are consistent with the SAXS results that AED76 can load 2 wt% of PEGDA575 or PETA with single hexagonal phase formation. Significant non-birefringence is found in the AED76PL2 (AED/glycerol/PEGDA250 = 76/22/2, *w*/*w*) system, meaning the destabilization of mesophase interfaces when loading 2 wt% of PEGDA250.

^{13}C solid-state NMR can further detect the crosslinkers' effect on the interfacial stability of the AED76 system by studying the interactions between the molecules in the systems, as shown in Figure 4 [5]. Peaks 'a' and 'b' (the peak splitting is due to the spin/spin coupling between ^{13}C and ^{1}H, since ^{1}H decoupling was not possible during the experiment) represent carbons in the reactive alkenyl group, while peak 'c' corresponds to the carbon in the carbonyl group. It is interesting to find that all the peaks become narrower and more significant after adding either PETA or PEGDA575 crosslinker, indicating that the introduction of crosslinkers leads to more liquid-like and less ordered HLLC systems. Moreover, no significant peak intensity changes occurred when adding more than 2 wt% of crosslinkers for both crosslinker systems, indicating that the extra crosslinker amount of more than 2 wt% is partitioned in the isotropic phase and does not have more interactions with AED surfactants.

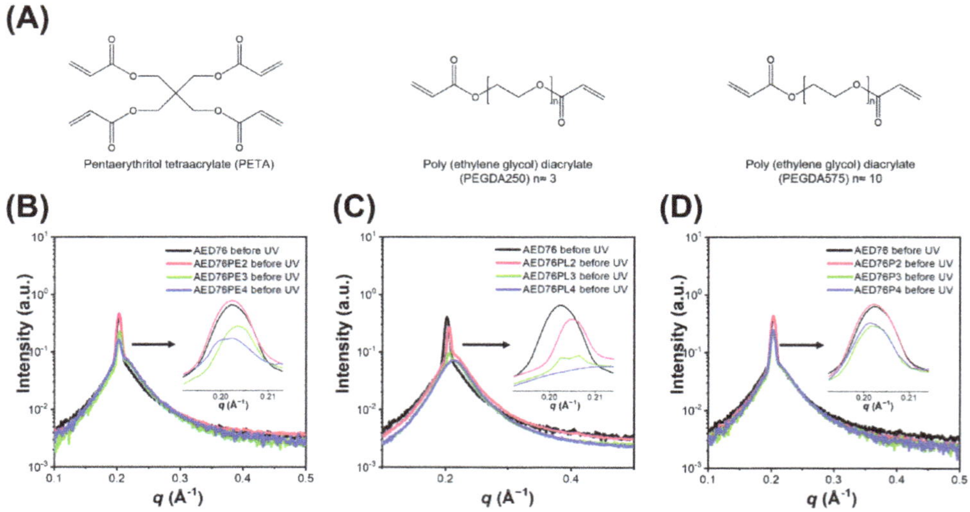

Figure 2. (**A**) Chemical structures of the crosslinkers: 1D SAXS pattern for AED76 system with crosslinker (**B**) PETA, (**C**) PEGDA250 and (**D**) PEGDA575. Inset: higher magnification of primary peak.

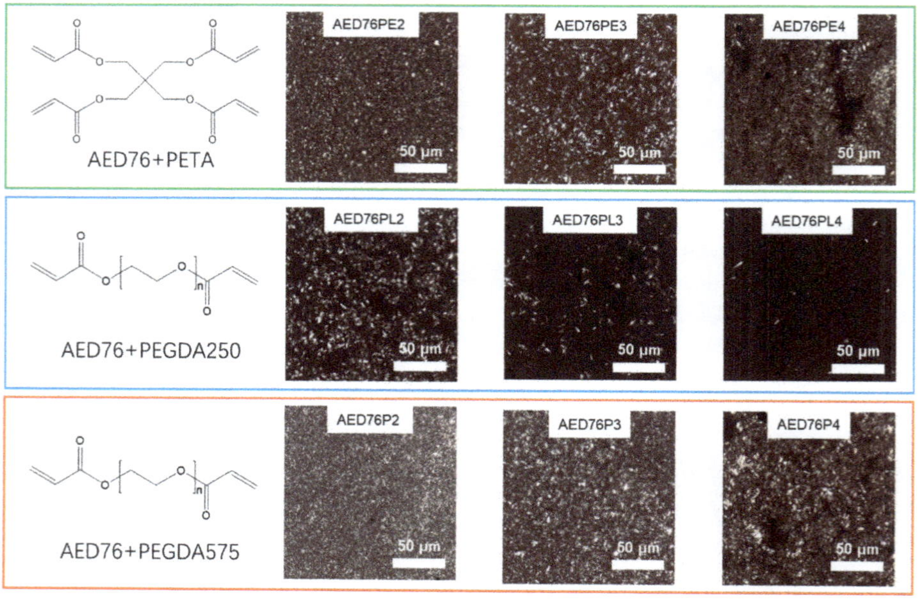

Figure 3. POM images for AED76 systems with crosslinkers PETA, PEGDA250 and PEGDA575 before polymerization.

Table 1. Dimensional elements of AED76, AED76P2 and AED76PE2 systems (D_{100}: lattice parameter, D_{inter}: the inter-channel distance between adjacent lipid cores, R: the radius of the lipid core, R_w: the thickness of water and head group, and S: the interfacial area per hydrophilic head group of the surfactant). The calculation of the parameters follows previously reported protocols [24].

Sample	Q_{100} (Å^{-1})	D_{100} (Å)	D_{inter} (Å)	R (Å)	R_w (Å)	S (Å^2)
AED76 before UV	0.2022	31.0583	35.8631	16.7495	2.3639	76.6915
AED76P2 before UV	0.2022	31.0583	35.8631	16.7313	2.4003	76.7750
AED76PE2 before UV	0.2022	31.0583	35.8631	16.3392	3.1847	78.6177

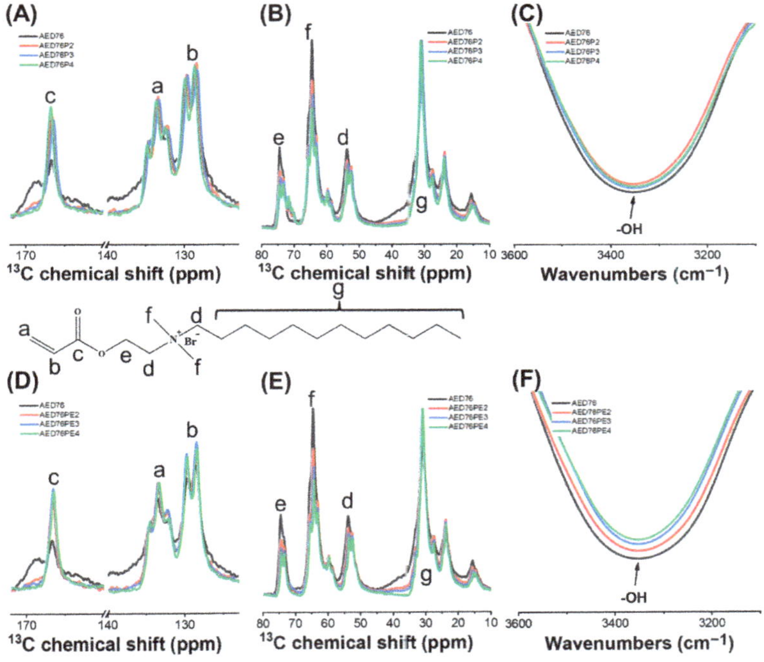

Figure 4. ^{13}C NMR spectra of AED76 systems before polymerization with (**A**) PEGDA575 and (**D**) PETA (chemical shifts from 123–172 ppm), and (**B**) PEGDA575 and (**E**) PETA (chemical shifts from 10–80 ppm); FTIR for AED76 systems with (**C**) PEGDA 575 and (**F**) PETA.

Additionally, it is seen that the intensity of peaks corresponding to the carbons around quaternary ammonium groups gradually decreases with the increasing concentration of crosslinkers, which is due to the decreasing concentration of glycerol after adding crosslinkers. Quaternary ammonium groups are in the hydrophilic/hydrophobic interfaces. The formation of hydrogen bonding among the quaternary ammonium groups of surfactants and the hydroxyl groups of the glycerol can de-shield the carbons near nitrogen atoms [25,26]. A lower concentration of glycerol leads to more significant chemical shielding of these carbons, which induces the gradual decrease in peak intensity.

FTIR measurements further confirm the interfacial stability of AED76 systems with various crosslinkers and indicate the positioning of the crosslinkers. Figure 4C,F present the FTIR spectra of AED76 systems with an increasing amount of PEGDA or PETA. The broad absorption band centered at ~3350 cm^{-1} arises from O-H stretching vibrations involving intra- and intermolecular hydrogen bonds between AED molecules and glycerol.

For the system with PEGDA, the peak area of -OH decreases (Figure S4) when adding 2 wt% of PEGDA into the AED76 system, indicating that the hydrogen bond interaction between glycerol and the head group of AED is weakened. It is interesting to find that the peak intensity of -OH becomes higher again after adding 3 or 4 wt% of PEGDA, indicating that the hydrogen bond interaction becomes stronger again. The peak intensity decreases and weakened hydrogen bond interaction can be attributed to the fact that the mesophase system retains its hexagonal structure when adding 2 wt% of PEGDA. When the mesophase becomes isotropic, the hydrogen bond interactions between the head group of AED and glycerol become stronger after adding more than 2 wt% of PEGDA.

In the systems with PETA, the peak intensity of -OH becomes lower with increasing concentration of PETA. Compared to PEGDA, hydrophobic PETA is more likely to be distributed close to hydrophilic/hydrophobic interfaces. The hydrogen bond interactions between the AED head groups and glycerol gradually become weaker, since the interactions between the PETA and AED become stronger with the increasing concentration of PETA. The interactions between the PETA and AED become even more significant in an isotropic state due to the lower steric hindrance.

3.2. The Role of Crosslinkers in the Retention of Hexagonal Structures and Other Properties of the Glycerol-Based HLLC-Templated Membranes

The introduction of structural optimized crosslinker enhances the retention of hexagonal structures, which improves the mechanical properties, transparency and water filtration performances of the glycerol-based HLLC-templated membranes.

The optimized interfacial stability of hexagonal mesophases increases the concentration accommodation of crosslinkers in AOE75H system, which results in enhanced retention of hexagonal structures and improved water filtration performances of the HLLC-templated membranes. From the SAXS pattern (Figure 5), the AED76P2 system after polymerization shows q_{100}, q_{110} and q_{200} (1:$3^{1/2}$:2) Bragg peaks corresponding to hexagonal phase, while the other two systems AED76 and AED76PE2 only show q_{100} and q_{200} (1:2) peaks. The significant q_{110} peak indicates that the AED76P2 can better retain the hexagonal structures after polymerization than AED76 and AED76PE2 [24,27,28]. The Gaussian fittings, center positions and FWHM of the scattering peaks are shown in Figure S5 and Table S3. High monomer conversion for AED76, AED76P2 and AED76PE2 during polymerization is shown in Figure S6.

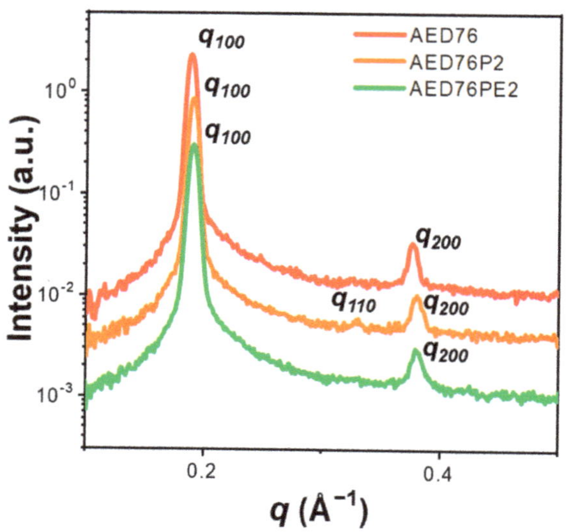

Figure 5. One-dimensional SAXS for AED76, AED76P2 and AED76PE2 systems after polymerization.

Figure 6A shows the crosslinkers' effect on the mechanical properties of the AED76 systems after polymerization. It is interesting to find that samples AED76 and AED76P2 have distinctly different stress/elongation curves. AED76 shows a low strength and very flexible behavior. AED76P2 also shows flexible behavior but a significant increase in mechanical strength due to the assistance of PEGDA575. The flexible behavior can be attributed to the plasticizing effect of glycerol, since there could be a small weight of glycerol remaining in the dead pores of both samples. The remaining glycerol in the membrane after polymerization and water swelling has been calculated (Section S2). If the glycerol is completely removed, the empty water channels in the membranes should be filled by water after water swelling. The experimental weight (0.0858 g) of the swelled AED76 system is apparently higher than the theoretical weight (0.0845 g), while these two weights are similar for the AED76P2 system (0.0741 vs. 0.0742 g). The higher experimental weight of the AED76 system indicates there is a considerable weight of unremoved glycerol in the system after water swelling, which is much less in AED76P2. This also suggests that the retained HLLC structure of the AED76P2 system contains more open pores. Furthermore, AED76P2 bulk material after polymerization becomes more transparent than AED76 (Figure 6B). The low visible light transmittance could be attributed to the phase rearrangement of the AED76 sample during UV-curing, which leads to different indexes of refraction in the system, while the better structure retention of the AED76P2 system results in uniform refraction indexes in the system.

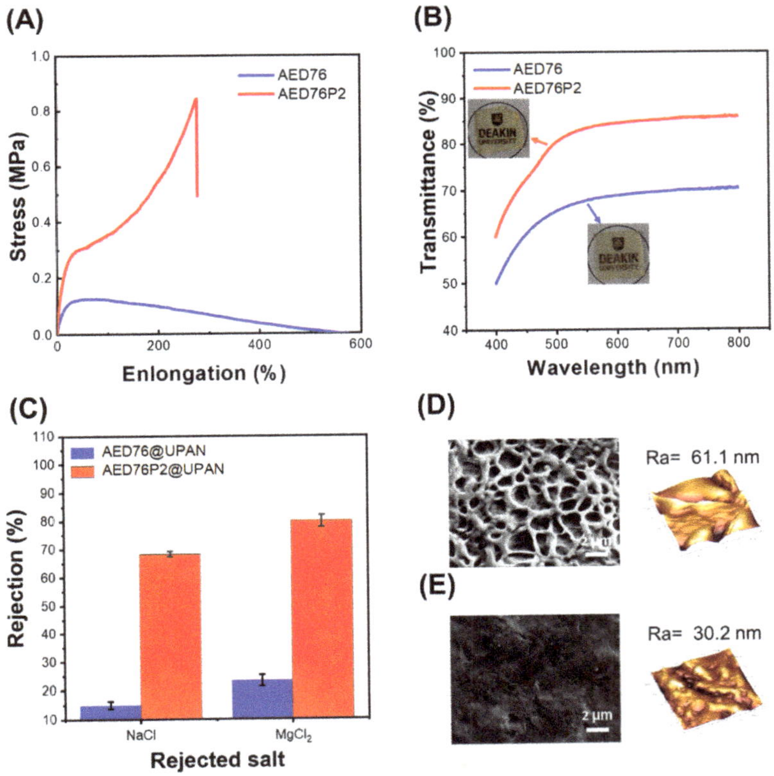

Figure 6. (**A**) Stress/elongation curve, (**B**) visible light transmittance pattern and (**C**) rejection performances for AED76 and AED76P2 systems; SEM images and AFM 3D height images for (**D**) AED76 system and (**E**) AED76P2 system.

The ion rejection performance of HLLC template active layers increased significantly after introducing the crosslinker, as shown in Figure 6C. The ion rejection rates of AED76P2@UPAN (NaCl: 68.3% and $MgCl_2$: 80%) are much larger than that of AED76@UPAN (NaCl: 14.9% and $MgCl_2$: 23.2%), which could be because the AED76P2 active layer (6.9 Å) has a much smaller estimated water pathway width than the AED76 active layer (13.09 Å) (Section S3). AED76P2 can theoretically reject all the hydrated Na^+ (7.2 Å). However, the soft nature of the AED76P2 material and the possible defects within the membrane could limit the rejection performance. The AED76 active layer still shows a 14.9% rejection rate of Na^+, although the estimated water pathway width surpasses the hydrated size of Na^+. This can be attributed to the electrostatic repulsion of quaternary ammonium cations. The thickness-normalized permeabilities of AED76@UPAN and AED76P2@UPAN are ~190 L m^{-2} h^{-1} bar^{-1} μm and ~40 L m^{-2} h^{-1} bar^{-1} μm (active layer thicknesses are shown in Figure S7), respectively. It is anticipated that the AED76P2@UPAN system can achieve a high water permeance of 200 L m^{-2} h^{-1} bar^{-1} when the active layer thickness is reduced to approximately 0.2 μm, which is significantly higher than the permeability typically found in commercial nanofiltration membranes (NF70-270, ~10–15 L m^{-2} h^{-1} bar^{-1}) [29,30]. This permeability also compares favorably to similar systems, such as 100 L m^{-2} h^{-1} bar^{-1} reported by Osuji et al. [2]. The high water permeability can be attributed to the crumpled membrane surface with high specific surface area [23,31,32] as shown in the Figure 6D,E. This surface morphology can be caused by the flexibility of the AED76P2 material. The surface roughness of the AED76 membrane (Ra = 61.1 nm) is twice that of AED76P2 (30.2 nm), which is attributed to the weaker crosslinked network of and higher unremoved glycerol in the AED76 system. Additionally, the hygroscopic effect of glycerol can also enhance the permeability of the systems [33]. This results in the much higher permeability of the AED76 system than AED76P2.

4. Conclusions

Previous studies have made significant advancements in using crosslinkers to assist the retention of reactive glycerol-based hexagonal structures [6,7]. Osuji et al. [7] found that glycerol-based hexagonal mesophase structures formed by mono-acrylate quaternary surfactants can be well retained through enhancing the chain entanglement at hydrophobic domains by incorporating a hydrophobic crosslinker. In this study, we investigated the effect of crosslinker structure on the interfacial stability of reactive glycerol-based HLLC mesophases as well as the concentration accommodation of crosslinkers within the mesophase through tuning the molecular length and hydrophilicity of crosslinkers. The direct effect originating from these variations in concentration accommodation on the retention of hexagonal structure and the water filtration performance of the glycerol-based HLLC-templated nanofiltration membranes were further studied.

Our key findings reaffirm the following:

- The poly (ethylene glycol) diacrylate (PEGDA) with ten ethylene glycol units exhibits high hydrophilicity, which minimizes the interference with the interfacial stability of hexagonal mesophases and enhances the concentration accommodation of crosslinkers.
- The hydrophilic PEGDA, with a long chain but low concentration of reactive groups, better connects the cylinders and retains the hexagonal structures compared to the hydrophobic one with shorter ethylene glycol units but a high concentration of reactive groups.
- The well-retained glycerol hexagonal structured nanofiltration membranes exhibit a remarkable pure water permeability of 40 L m^{-2} h^{-1} bar^{-1} μm, which is attributed to the strong hygroscopic effect of glycerol and the crumpled surface of membranes

due to the flexible nature of the system plasticized by glycerol. The permeability of the membranes is significantly higher than that of commercial membranes [29,30] and compares favorably to similar systems [2].

This work provides critical mechanistic insights into how crosslinker architecture governs interfacial stabilization in glycerol-derived hexagonal mesophases, while demonstrating a rational molecular design pathway for optimizing HLLC-templated nanofiltration membranes.

Supplementary Materials: The following supporting information can be downloaded at: https://www.mdpi.com/article/10.3390/cryst15070664/s1, Table S1: Sample compositions for studying binary AOE systems and ternary AOE systems with PEGDA 575; Section S1: Synthesis; Figure S1: the NMR spectra of the AED in d-chloroform ($CDCl_3$); Section S2: The calculation of the remained glycerol weight in AED samples after polymerization; Table S2: Weight variation of AED76 and AED76P2 membranes before and after UV-curing, and after water swelling; Section S3: Calculation of the pore dimensions in lamellar and hexagonal membranes; Figure S2: Schematic illustration for the (A) lamellar and (B) hexagonal pore dimensions; Figure S3: (A) 1D SAXS patterns and (B) POM images for AED/H2O binary systems; Figure S4: The integration area of -OH peaks; Figure S5: Peaks Gaussian fitting for AED76, AED76P2 and AED76PE2 systems after UV; Table S3: SAXS pattern peak center position and peak FWHM of AED76, AED76P2 and AED76PE2 systems after UV; Figure S6: FTIR spectra for AED76, AED76P2 and AED76PE2 before and after UV-curing; Figure S7: Cross-sectional surfaces for (A) AED76 and (B) AED76P2 active layers.

Author Contributions: Conceptualization, S.G.; methodology, S.G. and L.A.O.; validation, L.A.O.; formal analysis, S.G.; investigation, S.G.; resources, L.A.O.; data curation, S.G.; writing—original draft preparation, S.G.; writing—review and editing, S.G.; visualization, S.G.; supervision, L.K.; project administration, L.K.; funding acquisition, L.K. All authors have read and agreed to the published version of the manuscript.

Funding: The authors gratefully acknowledge the financial support of the Deakin University Postgraduate Scholarship and the financial support from the Australian Research Council (ARC) Discovery scheme (DP220102729). The authors thank Guang Wang from Chinese Spallation Neutron Source (CSNS) and Liliana de Campo from the Australian Neutron Science and Technology Organization (ANSTO), for their advice and help.

Data Availability Statement: The data are contained within the article.

Conflicts of Interest: The authors declare that there are no conflicts of interest regarding the publication of this manuscript.

References

1. Zhang, Y.; Kim, D.; Dong, R.; Feng, X.; Osuji, C.O. Tunable organic solvent nanofiltration in self-assembled membranes at the sub–1 nm scale. *Sci. Adv.* **2022**, *8*, eabm5899. [CrossRef] [PubMed]
2. Feng, X.; Imran, Q.; Zhang, Y.; Sixdenier, L.; Lu, X.; Kaufman, G.; Gabinet, U.; Kawabata, K.; Elimelech, M.; Osuji, C.O. Precise nanofiltration in a fouling-resistant self-assembled membrane with water-continuous transport pathways. *Sci. Adv.* **2019**, *5*, eaav9308. [CrossRef] [PubMed]
3. Dierking, I.; Martins Figueiredo Neto, A. Novel trends in lyotropic liquid crystals. *Crystals* **2020**, *10*, 604. [CrossRef]
4. Gu, S.; Yuan, B.; Bai, B.; Tong, X.; O'Dell, L.A.; Wang, D.; Kong, L.; Wang, G. Towards a high-flux separation layer from hexagonal lyotropic liquid crystals for thin-film composite membranes. *Membranes* **2021**, *11*, 842. [CrossRef] [PubMed]
5. Gu, S.; de Campo, L.; O'Dell, L.A.; Zhang, L.; Zhang, J.; Knott, R.; Zhang, J.; Yang, J.; Lynch, P.A.; Li, Y. Maintaining hexagonal structures through interfacial positioning of crosslinkers for nanofiltration. *J. Colloid Interface Sci.* **2025**, *683*, 182–192. [CrossRef] [PubMed]
6. Saadat, Y.; Tabatabaei, S.M.; Kim, K.; Foudazi, R. Thermoresponsive antifouling ultrafiltration membranes from mesophase templating. *J. Membr. Sci.* **2023**, *684*, 121861. [CrossRef]

7. Zhang, Y.; Dong, R.; Gabinet, U.R.; Poling-Skutvik, R.; Kim, N.K.; Lee, C.; Imran, O.Q.; Feng, X.; Osuji, C.O. Rapid fabrication by lyotropic self-assembly of thin nanofiltration membranes with uniform 1 nanometer pores. *ACS Nano* **2021**, *15*, 8192–8203. [CrossRef] [PubMed]
8. Wang, Y.; Li, Y.-X.; Li, Q.; Jia, R.; Tang, Q.; Huang, H.; Zhang, Y.; Feng, X. Highly Ordered Gyroid Nanostructured Polymers. Facile Fabrication by Polymerizable Pluronic Surfactants. *ACS Macro Lett.* **2024**, *13*, 550–557. [CrossRef] [PubMed]
9. Carter, B.M.; Wiesenauer, B.R.; Hatakeyama, E.S.; Barton, J.L.; Noble, R.D.; Gin, D.L. Glycerol-based bicontinuous cubic lyotropic liquid crystal monomer system for the fabrication of thin-film membranes with uniform nanopores. *Chem. Mater.* **2012**, *24*, 4005–4007. [CrossRef]
10. Auvray, X.; Petipas, C.; Anthore, R.; Rico, I.; Lattes, A. X-ray diffraction study of mesophases of cetyltrimethylammonium bromide in water, formamide, and glycerol. *J. Phys. Chem.* **2002**, *93*, 7458–7464. [CrossRef]
11. Auvray, X.; Perche, T.; Anthore, R.; Petipas, C.; Rico, I.; Lattes, A. Structure of Lyotropic Phases Formed by Sodium Dodecyl-Sulfate in Polar-Solvents. *Langmuir* **1991**, *7*, 2385–2393. [CrossRef]
12. Dörfler, H.; Senst, A. Influence of glycerol on the formation of lyotropic mesophases—Microscopic texture observations for determining preliminary phase diagrams of binary K-soap/glycerol systems. *Colloid Polym. Sci.* **1993**, *271*, 173–189. [CrossRef]
13. Li, P.; Reinhardt, M.I.; Dyer, S.S.; Moore, K.E.; Imran, O.Q.; Gin, D.L. Effects of structural modification of (alkyldiene-imidazolium bromide)-based gemini monomers on the formation of the lyotropic bicontinuous cubic phase. *Soft Matter* **2021**, *17*, 9259–9263. [CrossRef] [PubMed]
14. Robertson, L.A.; Schenkel, M.R.; Wiesenauer, B.R.; Gin, D.L. Alkyl-bis(imidazolium) salts: A new amphiphile platform that forms thermotropic and non-aqueous lyotropic bicontinuous cubic phases. *Chem. Commun.* **2013**, *49*, 9407–9409. [CrossRef] [PubMed]
15. Qian, Z.H.; Yue, X.; Yi, S.J.; Li, Q.T.; Chen, X. Unique lamellar lyotropic liquid crystal phases of nonionic phytosterol ethoxylates in glycerol. *RSC Adv.* **2015**, *5*, 101393–101400. [CrossRef]
16. Shukla, R.K.; Raina, K.K. Effect of viscosity, pH and physicochemical parameters of solvent on the aggregation and dielectric behaviour of lyotropic liquid crystals binary mixtures. *J. Mol. Liq.* **2018**, *250*, 71–79. [CrossRef]
17. Atkin, R.; Bobillier, S.M.C.; Warr, G.G. Propylammonium Nitrate as a Solvent for Amphiphile Self-Assembly into Micelles, Lyotropic Liquid Crystals, and Microemulsions. *J. Phys. Chem. B* **2010**, *114*, 1350–1360. [CrossRef] [PubMed]
18. Guàrdia, E.; Martí, J.; Padró, J.A.; Saiz, L.; Komolkin, A.V. Dynamics in hydrogen bonded liquids: Water and alcohols. *J. Mol. Liq.* **2002**, *96–97*, 3–17. [CrossRef]
19. Hajizadeh, M.; Golub, M.; Bektas, I.; Rusevich, L.L.; Embs, J.P.; Lohstroh, W.; Paulsen, H.; Pieper, J. Modulation of Protein Dynamics by Glycerol in Water-Soluble Chlorophyll-Binding Protein (WSCP). *Crystals* **2025**, *15*, 569. [CrossRef]
20. Saadat, Y.; Imran, O.Q.; Osuji, C.O.; Foudazi, R. Lyotropic liquid crystals as templates for advanced materials. *J. Mater. Chem. A* **2021**, *9*, 21607–21658. [CrossRef]
21. Arora, B.; Tandon, R.; Attri, P.; Bhatia, R. Chemical crosslinking: Role in protein and peptide science. *Curr. Protein Pept. Sci.* **2017**, *18*, 946–955. [CrossRef] [PubMed]
22. Zhang, J.; Zhang, S.; Liu, C.; Lu, Z.; Li, M.; Hurren, C.; Wang, D. Photopolymerized multifunctional sodium alginate-based hydrogel for antibacterial and coagulation dressings. *Int. J. Biol. Macromol.* **2024**, *260*, 129428. [CrossRef] [PubMed]
23. Wang, Z.; Wang, Z.; Lin, S.; Jin, H.; Gao, S.; Zhu, Y.; Jin, J. Nanoparticle-templated nanofiltration membranes for ultrahigh performance desalination. *Nat. Commun.* **2018**, *9*, 2004. [CrossRef] [PubMed]
24. Wang, G.; Garvey, C.J.; Zhang, J.; O'Dell, L.A.; Krause-Heuer, A.M.; Forsyth, M.; Darwish, T.A.; Miloš, S.; Kong, L. Evolution of structural dimensions in mesoporous template precursor from hexagonal lyotropic liquid crystals. *J. Phys. Condens. Matter* **2019**, *32*, 075101. [CrossRef] [PubMed]
25. Blackburn, J.C.; Kilpatrick, P.K. Using deuterium NMR lineshapes to analyze lyotropic liquid crystalline phase transitions. *Langmuir* **1992**, *8*, 1679–1687. [CrossRef]
26. Firouzi, A.; Atef, F.; Oertli, A.; Stucky, G.; Chmelka, B. Alkaline lyotropic silicate—Surfactant liquid crystals. *J. Am. Chem. Soc.* **1997**, *119*, 3596–3610. [CrossRef]
27. Zhang, J.; Xie, Z.; Hill, A.J.; She, F.H.; Thornton, A.W.; Hoang, M.; Kong, L.X. Structure retention in cross-linked poly (ethylene glycol) diacrylate hydrogel templated from a hexagonal lyotropic liquid crystal by controlling the surface tension. *Soft Matter* **2012**, *8*, 2087–2094. [CrossRef]
28. Gu, S.; Zhang, L.; de Campo, L.; Knott, R.; O'Dell, L.A.; Zhang, J.; Liu, K.; Li, X.; Yang, J.; Lynch, P.A. Interfacial design of support substrate for a continuous mesophase-templated active layer with adjustable pore size. *Colloids Surf. A* **2024**, *698*, 134569. [CrossRef]
29. Kéba Diawara, C.; Paugam, L.; Pontié, M.; Pierre Schlumpf, J.; Jaouen, P.; Quéméneur, F. Influence of chloride, nitrate, and sulphate on the removal of fluoride ions by using nanofiltration membranes. *Sep. Sci. Technol.* **2005**, *40*, 3339–3347. [CrossRef]
30. Cooper, J.; Ye, Y.; Razmjou, A.; Chen, V. High-Value Organic Acid Recovery from First-Generation Bioethanol Dunder Using Nanofiltration. *Ind. Eng. Chem. Res.* **2020**, *59*, 11940–11952. [CrossRef]

31. Li, X.; Wang, K.Y.; Helmer, B.; Chung, T.-S. Thin-film composite membranes and formation mechanism of thin-film layers on hydrophilic cellulose acetate propionate substrates for forward osmosis processes. *Ind. Eng. Chem. Res.* **2012**, *51*, 10039–10050. [CrossRef]
32. Maruf, S.H.; Greenberg, A.R.; Pellegrino, J.; Ding, Y. Fabrication and characterization of a surface-patterned thin film composite membrane. *J. Membr. Sci.* **2014**, *452*, 11–19. [CrossRef]
33. Mondal, S.M.; Chaudhary, A. *Sustainable Development of an Ultrafiltration Membrane from Banana Pseudo Stem for Use in Water Filtration*; Jaypee University of Information Technology: Solan, India, 2023.

crystals

Article

The Effect of Nematic Liquid Crystal on the Performance of Dye-Sensitized Solar Cells

Paweł Szubert and Stanisław A. Różański *

Department of Electrical Engineering, Stanisław Staszic State University of Applied Sciences in Piła, 64-920 Piła, Poland; pszubert@ans.pila.pl
* Correspondence: srozansk@asta-net.com.pl

Abstract

The motivation for increasing the efficiency of renewable energy sources is the basic problem of ongoing research. Currently, intensive research is underway in technology based on the use of dye-sensitized solar cells (DSSCs). The aim of this work is to investigate the effect of modifying the iodide electrolyte with liquid crystals (LCs) known for the self-organization of molecules into specific mesophases. The current–voltage (I-V) and power–voltage (P-V) characteristics were determined for the ruthenium-based dyes N3, Z907, and N719 to investigate the influence of their structure and concentration on the efficiency of DSSCs. The addition of a nematic LC of 4-*n*-pentyl-4-cyanobiphenyl (5CB) to the iodide electrolyte influences the I-V and P-V characteristics. A modification of the I-V characteristics was found, especially a change in the values of short circuit current (I_{SC}) and open circuit voltage (V_{OC}). The conversion efficiency for cells with modified electrolyte shows a complex dependence that first increases and then decreases with increasing LC concentration. It may be caused by the orientational interaction of LC molecules with the titanium dioxide (TiO_2) layer on the photoanode. A too high concentration of LC may lead to a reduction in total ionic conductivity due to the insulating effect of the elongated polar molecules.

Keywords: nematic liquid crystal; I-V characteristics; short circuit current; open circuit voltage; dye-sensitized solar cells; electrolyte

1. Introduction

Renewable energy sources have become particularly important in the era of problems with access to energy produced from non-renewable sources [1]. Solar energy is usually converted into an electric current using photovoltaic cells based on semiconductor properties [2]. They are classified as first-generation photovoltaic cells, which use monocrystalline silicon produced by the Czochralski method or polycrystalline silicon, which is obtained by melting and appropriately forming silicon [3,4]. The efficiency of these cells is in the range of 18–25.6% [5]. The second-generation cells are thin-film systems (layer thickness 1–3 μm) built from monolayers of cadmium telluride (CdTe), a mixture of copper, indium, gallium, and selenium (CIGS), or amorphous silicon (a-Si). The efficiency changes from about 8% for a-Si cells to 17% for CdTe modules and 16% for CIGS [6]. In addition, third-generation photovoltaic cells based on a reversible photochemical process, known as dye-sensitized solar cells (DSSCs), are being intensively studied [7–11].

LCs, due to their intermediate properties between liquids and solid crystals and the ability of molecules to self-organize into variously ordered mesophases, are finding

 https://doi.org/10.3390/cryst15080705

increasing applications not only in information imaging devices [12–16]. In particular, their unique optical and electro-optical properties and ease of deformation and reorientation by means of electric, magnetic, mechanical, and other external fields arouse constant interest in these materials. In many modern devices designed for energy conversion or storage—such as DSSCs, fuel cells, Li-ion batteries, and supercapacitors—an organic electrolyte is a key component. Its stability and physicochemical properties directly translate into the efficiency of these systems [17–25]. The essence of the concept of using LCs is that properly ordered structures on the nanometer scale contain an ionophilic and ionophobic part, leading to two functions, namely the formation of conductive channels and insulating moieties. Moreover, insulating parts fulfill the role of stabilizing the system mechanically and thermally [16]. Other advantages of using LCs include their ability to self-organize into systems with anisotropic properties, leading to an increase in ionic conductivity in a specific direction. Moreover, temperature induction of phase transitions enables switching of functions in a given electrolytic structure [13].

The aforementioned LC properties are employed in the design and synthesis of functional materials. They provide high ionic conductivity of electrolytes while maintaining safety, chemical resistance, and electrochemical stability. Additionally, they enhance the mechanical robustness of devices and help lower production costs [20]. A promising class of compounds is ionic liquid crystals (ILCs) that combine the properties of ionic liquids and LCs. ILCs consist of rigid aromatic molecular cores influencing long-range ordering and ionic centers and alkyl chains enabling modulation of charge mobility [20,21,24,26–28]. Long alkyl chains and spontaneous self-organization processes favor the formation of ordered mesophases on the macroscopic scale. Consequently, the system most often transitions into a smectic phase. Applications of ILCs involve the use of the organized, anisotropic, and partially ordered structure of ionic liquids to increase the conductivity of matter and charge, e.g., in DSSCs. For example, a binary electrolyte composed of a carbonate-functionalized mesogen and an iodide ionic liquid nanosegregates into lamellar structures exhibits more than four times higher ionic conductivity parallel rather than perpendicular to the smectic layers. By using this method, a thermally stable and durable LC-electrolyte is obtained, enabling the redox reaction I^-/I_3^- over a wide temperature range and higher conversion rates. This can be attributed to longer electron lifetimes and higher electron densities in the photoelectrodes [29,30]. The I^-/I_3^- redox couple plays a crucial role in regenerating photoexcited dye molecules in DSSCs. The process includes oxidizing iodide (I^-) to triiodide (I_3^-) by the oxidized dye molecules. The triiodide then diffuses to the counter electrode, where it is reduced back to iodide. Moreover, a method based on doping the electrolyte with LC E7 [17] was developed. After adding LC to the electrolyte prepared on the basis of 3-methoxypropionitrile, no mesophase was detected in the resulting mixture. The results show that small amounts of LC reduce the short circuit current density for DSSC because LC reduces the rate of electrochemical reaction between the DSSC counter electrode and the electrolyte. However, the LC dopant delays the rate of DSSC degradation due to the interaction between cyano-groups and organic solvent in the liquid electrolyte. Due to molecular interactions, LCs increase the viscosity and stability of the electrolyte, thus inhibiting the rate of its evaporation. It has also been shown that mesogenic molecules can form strong coordination bonds with titanium atoms on the surface of the porous TiO_2 layer, improving the stability and photovoltaic efficiency of the DSSC cell [31].

The use of electrolytes based on volatile organic solvents in DSSCs raises great concerns related to their commercialization due to flammability, instability, evaporation, or leakage tendency. Alternative solutions based on solid or quasi-solid materials, such as composites, polymers, hole-conducting materials, and LCs, are attracting increasing attention, especially the latter, which are promising solutions as a charge transport layer in DSSCs.

Taking into account the above-mentioned properties of LCs in the context of the conducted research, the following questions should be answered: (i) How does the presence of LC molecules in DSSC cells affect their basic parameters such as short circuit current, open circuit voltage, fill factor, maximum power, or conversion efficiency? (ii) What is the role of LCs in electrolyte stabilization and durability of DSSC cells? The aim of this work is to provide at least a partial answer to the questions posed and problems indicated above.

2. Materials and Methods

2.1. DSSC Cell Preparation Procedure

The measurement of DSSC cells involves a multilayer system consisting of a photoanode (glass with a conductive layer coated with TiO_2 with an absorbed dye), an electrolyte, and a cathode coated with a platinum layer as a redox reaction catalyst (Figure 1). Most of the materials and reagents necessary for the construction of the test DSSC cells were purchased from Solaronix [32]. The measurement of DSSC cells involved using glass plates with a size of 20 × 20 mm covered with a transparent conductive layer of fluorine-doped tin oxide (FTO). In addition, the porous layer of TiO_2 with an area of about 6 × 6 mm was applied on the anode surface.

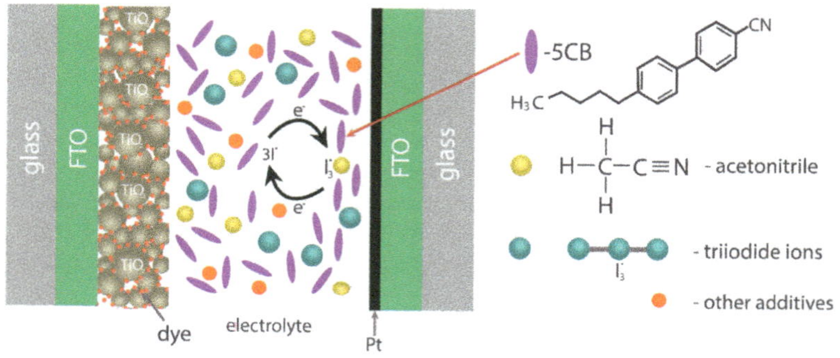

Figure 1. Schematic diagram of the DSSC cell structure with marked electrolyte components.

Preparation for the measurement of DSSC cells consisted of several stages. First, solutions of a specific concentration were prepared by dissolving the dyes in spectrally pure ethanol. The concentration is determined by weighing specific masses of the components of the solution. A Mettler Toledo balance was used for this purpose, and each solution was placed in a tightly closed dark glass bottle. Before closing the bottles, the solutions were mixed using a glass rod until the dye crystals were completely dissolved. The solutions prepared in this way were left for several hours to ensure that the dye had completely dissolved. Then, the dye solution prepared in this way was poured into a small Petri dish, and a previously prepared anode with a porous TiO_2 layer was placed in it. Before that, the anode should be heated to remove water and other impurities. The dish was sealed with wax paper and placed for 12 h in a shaded desiccator to absorb the dye (anode sensitization). The experience of other authors shows that the time of dye sensitization is important for the efficiency of DSSC cells [33,34].

After the sensitization process, the anode was removed, washed with ethanol, and dried to eliminate excess solvent with the dye. Then, a 60 μm thick Meltronix polymer film spacer (DuPont Surlyn®, Wilmington, DE, USA) (size: 14 × 14 mm; aperture: 8 × 8 mm) is applied to the anode with a porous TiO_2 layer to seal the system and prevent electrode short-circuiting. Sealing of the system occurred as a result of the melting of the gasket at a certain temperature (about 373 K) and pressure. The cathode was a glass plate covered with FTO and a platinum layer, which is a redox catalyst. A hole drilled in the cathode to fill the cell with electrolyte is positioned above the TiO_2 layer. The sandwich was assembled in such a way as to leave edges to which measuring electrodes can be connected. The resulting cell was filled with iodide electrolyte using capillary forces through a hole in the cathode. After filling the cell with electrolyte, the hole in the cathode was closed using a glass plate and Meltronix polymer film, which glues the system together under the influence of temperature. The prepared samples were stored in low humidity and in the absence of lighting. A dark glass desiccator with a hygroscopic insert was used.

2.2. Dyes and Electrolytes

As mentioned above, the DSSC with porous TiO_2 electrodes was sensitized using ruthenium-based dyes purchased at Solaronix [32]. In this experiment the following ruthenium dyes were used: cis-diisothiocyanato-bis (2,2′-bipyridil-4,4′-dicarboxylic acid), ruthenium (II) known as N3, cis-diisothiocyanato-bis (2,2′-bipyrida-4,4′-dicarboxylato), ruthenium (II) bis (tetrabutylammonium) marked as N719, and the amphiphilic ruthenium dye cis-diisothiocyanato-(2,2′-bipyridil-4,4′-dicarboxylic acid)-(2,2′-bipyridil-bipipyridil-4,4′-dinonyl) ruthenium (II) known in literature as Z907 (Figure 2). The spectral properties of N3, N719, and Z907 dyes and their use in the DSSC cells were investigated, e.g., in the works [35,36]. The low-viscosity electrolyte (iodolyte AN-50), contained as a redox couple iodide/triiodide and a redox concentration of 50 mM, was used. The electrolyte was prepared on the basis of acetonitrile solvent with the addition of ionic liquid, lithium salt, and pyridine derivative [32]. Acetonitrile is a low-reactive organic solvent with a relatively low viscosity of approximately 0.369 mPa·s at 298 K. The dipole moment of the acetonitrile molecule is about 3.92 D. The evaporation temperature of this solvent is 354.6 K. The electrolyte was stored in tight bottles in a fume hood. All manipulations were performed using sterile Pasteur pipettes.

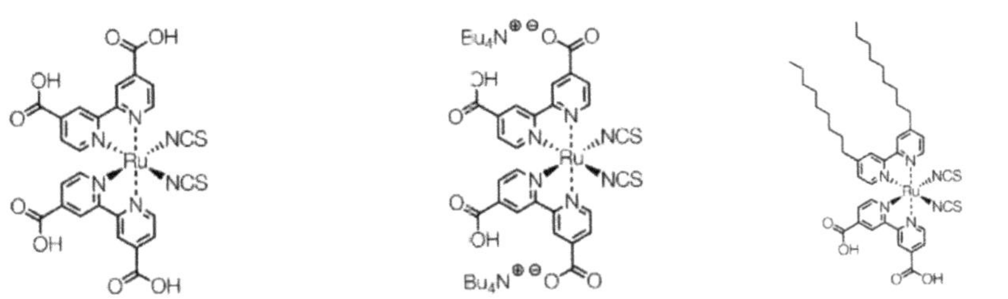

(a) $C_{26}H_{16}O_8N_6S_2Ru$ (b) $C_{58}H_{86}O_8N_8S_2Ru$ (c) $C_{42}H_{52}O_4N_6S_2Ru$

Figure 2. Chemical structures of ruthenium-based dyes: (**a**) N3; (**b**) N719; (**c**) Z907 [32].

2.3. Preparation of Samples with Liquid Crystal

The LC 5CB has a nematic phase below 308 K but crystallizes at 295.7 K. 5CB is formed by rod-like molecules with a rigid polarizable core of two phenyl rings with a cyano-group and a flexible aliphatic end group formed by a pentyl chain. The dipole moment of this polar molecule is approximately 4.8 D [37]. For this anisotropic material, the perpendicular viscosity is 129 mPa·s and the parallel viscosity is 45 mPa·s at 294 K [38]. In order to check the effect of 5CB on the operating parameters of the DSSC cells, several solutions of AN-50 iodide electrolyte (Solaronix, Aubonne, Switzerland) with an admixture of LC were prepared (10% wt, 15% wt). After weighing the appropriate portion of electrolyte and adding the LC to it, the solution was maintained at a temperature above 308 K (5CB isotropic phase) to obtain a homogeneous mixture. An ultrasonic washer with regulated water temperature was used for this purpose. The process lasted about 15 min at a temperature of about 313 K. Additionally, in order to obtain a homogeneous mixture, the solution was stirred using a glass rod. The DSSC cells were filled with the electrolyte prepared in this way as a result of capillary forces.

Solutions at 10% wt and 15% wt of 5CB in AN-50 electrolyte were examined under a polarizing microscope in both thin and thick cells. With crossed polarizers, the field of view was black. This suggests that the electrolyte prepared in this way behaves as an isotropic liquid without the presence of a nematic phase. The 5CB molecules dispersed in the electrolyte are incapable of forming a nematic phase at the concentrations discussed. However, polar LC molecules can interact with ions present in solution and influence their mobility.

2.4. The Current–Voltage Characteristics

Two Keysight 34465A multimeters were used to measure the current–voltage (I-V) characteristics of DSSC cells, along with dedicated Keysight BenchVue 2020 Update 1.0 software, which allows data collection in MS Excel files. Integration of the measurement system with the Origin 2023 (OriginLab) environment allowed for detailed analysis of the results and determination of basic cell parameters. The PET "Photo Emission Tech" sun simulator Cell Tester Model #CT150AAA was used as a source of light radiation. The optimal intensity of the simulator is 10^3 W/m^2. The load used in the tests was a resistor consisting of seven decades with an accuracy class of 0.05 Ω and the possibility of changing the load from 0 to 1 MΩ. The temperature control was applied at the sample measurement site. The measurements were carried out under conditions of stable temperature of about 305 K.

In the case of an ideal photovoltaic cell, the I-V characteristic should have the shape of a rectangle with coordinates I_{SC} and V_{OC} (Figure 3). The maximum power of such a cell, $P_0 = I_{SC}V_{OC}$, is always greater than the power of the real cell, $P_{max} = I_mV_m$, where I_m and V_m are the coordinates of the P_{max}. Figure 3 also illustrates how the P_{max} value can be determined from the power–voltage (P-V) characteristics.

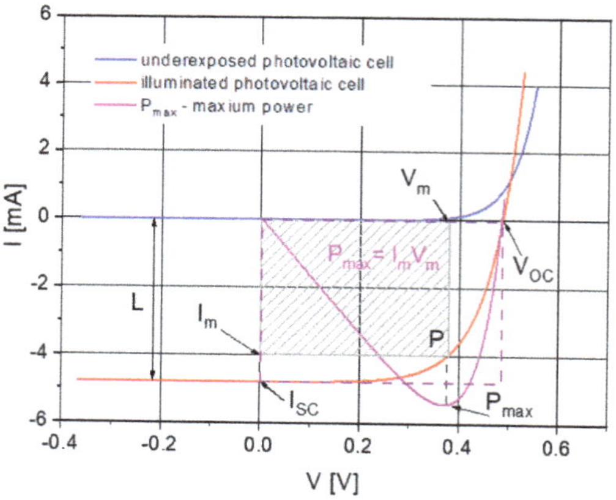

Figure 3. Diagram illustrating the method of deriving the photovoltaic parameters from the I-V characteristic.

The I-V characteristic of a real cell deviates from the rectangular shape and can be approximately described by the simplified dependency [39–41]:

$$I = I_{ph} - I_0\left[\exp\left(\frac{qV}{nk_BT}\right) - 1\right] \tag{1}$$

where I_{ph} represents solar cell current, I_0 is a diode saturation current, q is an electron charge (1.6×10^{-19} C), k_B is a Boltzmann constant (1.38×10^{-23} J/K), n is an ideality factor (from 1 to 2), and T is an absolute temperature. The second term in Equation (1) represents the known Shockley's diode current formula for semiconductor materials [42].

An important parameter for a PV cell is the efficiency factor η of converting light energy into electric current, which is defined as the ratio of the P_{max} to the power of light L incident on the cell surface S:

$$\eta = \frac{P_{max}}{LS} \times 100\% \quad (2)$$

The next important parameter is the fill factor FF, which is defined as follows:

$$FF = \frac{P_{max}}{I_{SC} V_{OC}} \quad (3)$$

By substituting (3) into (2), the nominal value of the conversion factor can be defined as follows:

$$\eta = \frac{I_{SC} V_{OC} FF}{LS} \times 100\% \quad (4)$$

Taking into account Equation (1) and using the definition of power, the power dissipated in the DSSC cell can be described by the following function:

$$P = V\left\{I_{ph} - I_0\left[\exp\left(\frac{qV}{nk_B T}\right) - 1\right]\right\} \quad (5)$$

The optimal load value is calculated using the following equation:

$$R = \frac{V_m}{I_m} \quad (6)$$

3. Results

Figure 4a–c show the I-V characteristics of DSSC cells sensitized by three different ruthenium-based dyes, namely N3, Z907, and N719, for selected light intensities. The characteristics were obtained for a dye concentration of about 0.05% wt in ethanol. The dependence of the I_{SC} on the light intensity is typical of photovoltaic cells—changes in I_{SC} are proportional to changes in the light intensity.

It can be seen that with the increase in light intensity, the V_{OC} slightly increases. The solid lines marked on the figures were obtained by fitting Equation (1) to the experimental points. Although this function describes the photovoltaic phenomenon in a rather simplified way, the obtained fits are rather good. Moreover, Figure 4d,e, and f show P-V characteristics of the investigated DSSC cells for the three mentioned above dyes. The experimental points are fitted with Equation (5) in order to find the P_{max} of the DSSC cells. In the nonlinear part of the I-V or P-V characteristics, the fits with Equations (1) and (5) deviate more from the measurement points for the higher light intensities illuminating the sample. The fitting parameters are summarized in Tables 1–3. The parameters such as I_{SC}, V_{OC}, P_0, and P_{max} increase with increasing light intensity, but their nominal values depend on the dye used. Interestingly, the FF parameter increases slightly with decreasing light intensity, whereas the efficiency η exhibits a maximum for L = 121 W/m^2. The observed effect of light intensity on the DSSC parameters is consistent with theoretical predictions and results of other experiments [43,44]. Moreover, the I_{ph} obtained from the fitting is in good accordance with the values of I_{SC} read from the I-V characteristics. However, the obtained value of the ideality factor n needs a special elucidation because it differs significantly from the theoretical value. The Shockley–Read–Hall recombination theory predicts that the value of n should be equal to or less than 2. One explanation for this deviation is that the recombination current does not flow uniformly throughout the cell but always in local sites, e.g., in extended defects [45].

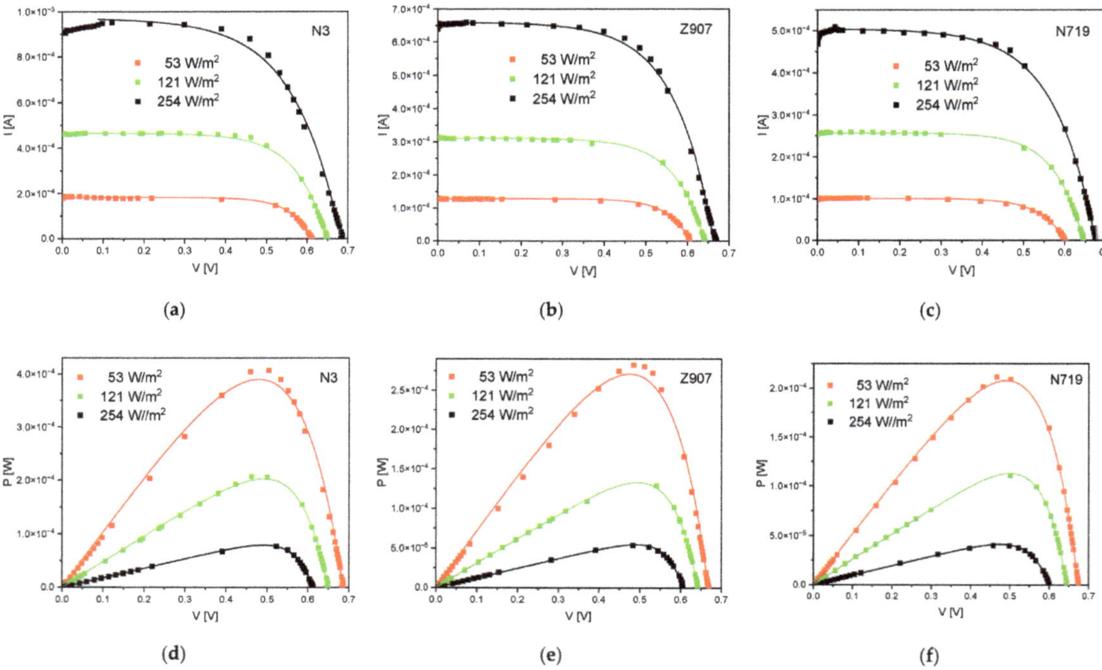

Figure 4. Comparison of the I-V characteristics for different dyes at chosen light illuminations: (**a**) N3, (**b**) Z907, and (**c**) N719. Comparison of the P-V characteristics for different dyes: (**d**) N3, (**e**) Z907, and (**f**) N719.

Table 1. The DSSC cell parameters for N3 dye.

	N3		
L [W/m^2]	53	121	254
I_{SC} [A]	1.85×10^{-4}	4.63×10^{-4}	9.69×10^{-4}
V_{OC} [V]	0.61	0.65	0.69
P_0 [W]	1.13×10^{-4}	3.01×10^{-4}	6.69×10^{-4}
P_{max} [W]	0.80×10^{-4}	2.05×10^{-4}	4.07×10^{-4}
FF	0.71	0.68	0.61
η [%]	4.19	4.70	4.43
I_{ph} [A]	$(1.826 \pm 0.004) \times 10^{-4}$	$(4.64 \pm 0.01) \times 10^{-4}$	$(9.69 \pm 0.11) \times 10^{-4}$
I_0 [A]	$(2.19 \pm 0.37) \times 10^{-9}$	$(7.33 \pm 1.27) \times 10^{-8}$	$(2.39 \pm 0.58) \times 10^{-6}$
n	18.5 ± 0.3	13.5 ± 0.3	8.8 ± 0.4

Figure 5a shows a comparison of I-V characteristics for different dyes at a selected constant sample illumination. The highest I_{SC} is shown by the sample sensitized with N3 dye, then with Z907 dye, and finally with N719 dye. Additionally, a slight shift of V_{OC} towards higher values is observed. The efficiency of DSSC cells depends to a large extent on the photosensitive dyes used and, in particular, on their molecular structure and the location of the HOMO/LUMO energy levels in relation to the energy levels of the TiO_2 semiconductor applied to the anode. The N3 and N719 dye molecules are quite large and compact molecular systems differing from each other by the tetrabutylammonium groups (Figure 2). However, the Z907 dye molecule has a structure similar to the N3 but has two long alkyl chains attached, which makes it an amphiphilic molecule. The discussed modifications of ruthenium-based molecules can affect the efficiency of DSSC cells sensitized with these molecules. On the other hand, Figure 5b shows the dependence of power dissipated in the DSSC cell on the voltage for the three dyes mentioned above.

Table 2. The DSSC cell parameters for N719 dye.

	N719		
L [W/m^2]	53	121	254
I_{SC} [A]	1.00×10^{-4}	2.56×10^{-4}	5.04×10^{-4}
V_{OC} [V]	0.60	0.65	0.67
P_0 [W]	0.60×10^{-4}	1.66×10^{-4}	3.40×10^{-4}
P_{max} [W]	0.41×10^{-4}	1.12×10^{-4}	2.11×10^{-4}
FF	0.68	0.67	0.62
η [%]	2.15	2.57	2.30
I_{ph} [A]	$(1.004 \pm 0.002) \times 10^{-4}$	$(2.560 \pm 0.003) \times 10^{-4}$	$(5.04 \pm 0.02) \times 10^{-4}$
I_0 [A]	$(4.84 \pm 0.66) \times 10^{-9}$	$(2.00 \pm 0.17) \times 10^{-8}$	$(5.90 \pm 0.68) \times 10^{-7}$
n	16.5 ± 0.2	14.7 ± 0.1	10.0 ± 0.2

Table 3. The DSSC cell parameters for Z907 dye.

	Z907		
L [W/m^2]	53	121	254
I_{SC} [A]	1.29×10^{-4}	3.10×10^{-4}	6.57×10^{-4}
V_{OC} [V]	0.61	0.64	0.67
P_0 [W]	0.79×10^{-4}	1.98×10^{-4}	4.40×10^{-4}
P_{max} [W]	0.54×10^{-4}	1.32×10^{-4}	2.82×10^{-4}
FF	0.68	0.67	0.64
η [%]	2.83	3.03	3.07
I_{ph} [A]	$(1.278 \pm 0.003) \times 10^{-4}$	$(3.105 \pm 0.003) \times 10^{-4}$	$(6.59 \pm 0.03) \times 10^{-4}$
I_0 [A]	$(1.24 \pm 0.27) \times 10^{-9}$	$(4.81 \pm 0.34) \times 10^{-8}$	$(5.27 \pm 0.96) \times 10^{-7}$
n	19.04 ± 0.4	13.7 ± 0.1	10.7 ± 0.3

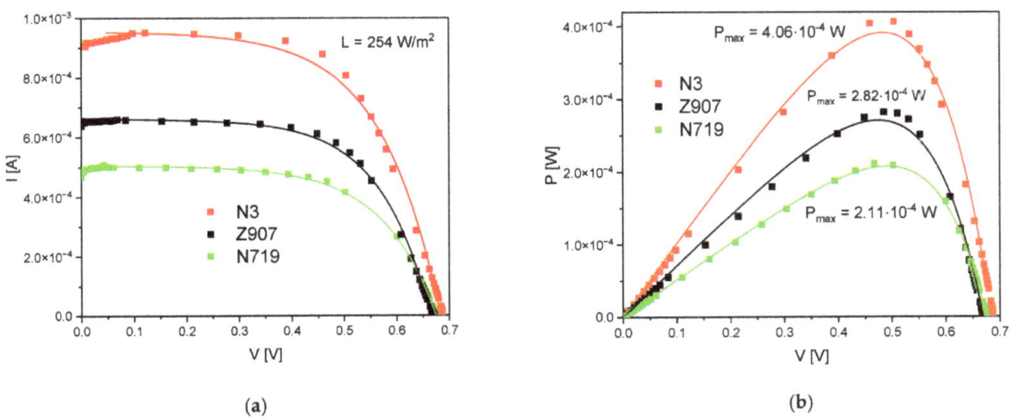

Figure 5. (**a**) Comparison of I-V characteristics of the DSSC cells with different dyes; (**b**) P-V characteristics for determination of the maximum power.

The solid lines marked in the figures were obtained as before by fitting the measurement results with Equations (1) and (5). This procedure allows for determining the maximum power as well as other basic parameters of DSSC cells, which are listed in Table 4. As can be seen from Table 4, almost all parameters of DSSC cells sensitized with dyes in the order of N3, Z907, and N719 decrease. However, it is worth noting that the V_{OC} for the cell with the Z907 dye lies between the values for cells with N3 and N719 dyes. This leads to a situation where the *FF* for the cell with the Z907 dye is the highest. It should be noted that the efficiency of the cells is not very high and varies from 4.4% to 2.3%.

Table 4. Comparison of the DSSC cells' parameters for different dyes (L = 254 W/m^2).

Sample	I_{SC} [A]	V_{OC} [V]	P_0 [W]	P_{max} [W]	*FF*	η [%]	I_{ph} [A]	I_0 [A]	n
N3	9.54×10^{-4}	0.687	6.55×10^{-4}	4.07×10^{-4}	0.62	4.4	$(9.69 \pm 0.11) \times 10^{-4}$	$(2.39 \pm 0.58) \times 10^{-6}$	8.8 ± 0.4
Z907	6.60×10^{-4}	0.665	4.39×10^{-4}	2.82×10^{-4}	0.64	3.1	$(6.59 \pm 0.03) \times 10^{-4}$	$(5.27 \pm 0.96) \times 10^{-7}$	10.7 ± 0.3
N719	5.04×10^{-4}	0.674	3.40×10^{-4}	2.11×10^{-4}	0.62	2.3	$(5.04 \pm 0.02) \times 10^{-4}$	$(5.90 \pm 0.68) \times 10^{-7}$	10.0 ± 0.2

Figure 6 presents a summary of measurements taken for DSSC cells containing ruthenium N3 dye at different concentrations. As can be seen in Figure 6a, with the increase in dye concentration for a given light intensity, the I_{SC} increases, while the V_{OC} does not change much. However, the shift in V_{OC} for higher voltages is observed for 0.07% wt concentration. On this basis, it can be concluded that the efficiency of the cell increases with the increase in dye concentration. This relationship seems to be nonlinear. This could be expected because increasing the amount of dye in the porous TiO_2 anode increases the efficiency of converting light into electric current—a larger number of dye molecules are in contact with the semiconductor surface, which facilitates the release of electrons to the external circuit. It should be noted that despite small changes in dye concentration, the changes in the value of the I_{SC}, especially for the concentration of 0.09% wt, seem to be significant.

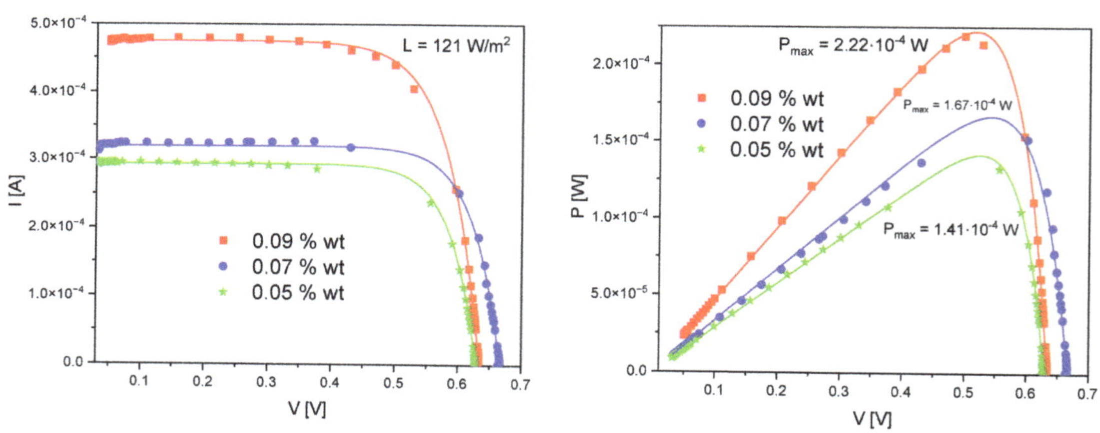

(**a**) (**b**)

Figure 6. (**a**) I-V characteristics of the DSSC cell for different N3 dye concentrations; (**b**) P-V characteristics of DSSC cells for different N3 dye concentrations.

It should be noted that for sensitization of DSSC cells, it is unfavorable to use too low or too high dye concentrations, which is associated with too low a level of light conversion in the first case or increased agglomeration of the dye and reduced its efficiency in the second case [35,46].

Table 5 shows the results obtained by fitting the experimental data with Equations (1) and (5) for three selected N3 dye concentrations. As can be seen, the P_{max}, I_{SC}, and η increase with the increase in dye concentration. However, V_{OC} and FF change slightly. Moreover, the above parameters somewhat differ from the observed trend for the 0.07% wt concentration, especially for V_{OC} and FF.

Table 5. Parameters of the DSSC cells for different N3 dye concentrations (L = 121 W/m^2).

N3 Dye [% wt]	I_{SC} [A]	V_{OC} [V]	P_0 [W]	P_{max} [W]	FF	η [%]	I_{ph} [A]	I_0 [A]	n
0.05	2.93×10^{-4}	0.627	1.84×10^{-4}	1.41×10^{-4}	0.77	3.2	$(2.932 \pm 0.004) \times 10^{-4}$	$(4.42 \pm 0.74) \times 10^{-11}$	25.0 ± 0.3
0.07	3.20×10^{-4}	0.665	3.16×10^{-4}	1.67×10^{-4}	0.53	3.8	$(3.199 \pm 0.009) \times 10^{-4}$	$(4.75 \pm 1.76) \times 10^{-11}$	23.6 ± 0.5
0.09	4.75×10^{-4}	0.635	3.02×10^{-4}	2.22×10^{-4}	0.74	5.1	$(4.750 \pm 0.008) \times 10^{-4}$	$(1.13 \pm 0.23) \times 10^{-9}$	20.5 ± 0.3

Figure 7a presents the I-V characteristics for a DSSC cell with an N3 dye at a weight concentration of about 0.05% wt. The characteristics of three DSSC cells differing in the preparation of the electrolyte were compared. The results marked with asterisks were obtained for cells filled with unmodified AN-50 electrolyte, while the remaining results refer to DSSC cells with an electrolyte modified with 5CB at different concentrations. The obtained characteristics show a rather complex and unexpected course. Figure 7b shows the power dependence on the voltage for various concentrations of 5CB in the electrolyte. Based on the presented relationship, P_{max} for the compared DSSC cells was determined. It can be noticed that the P_{max} of the DSSC cell with electrolyte without an LC is the smallest, while the P_{max} of the cell with a lower concentration of 5CB is higher compared to the other two. Based on the characteristics shown in Figure 7, the basic parameters of the studied DSSC cells were determined. The results are presented in Table 6.

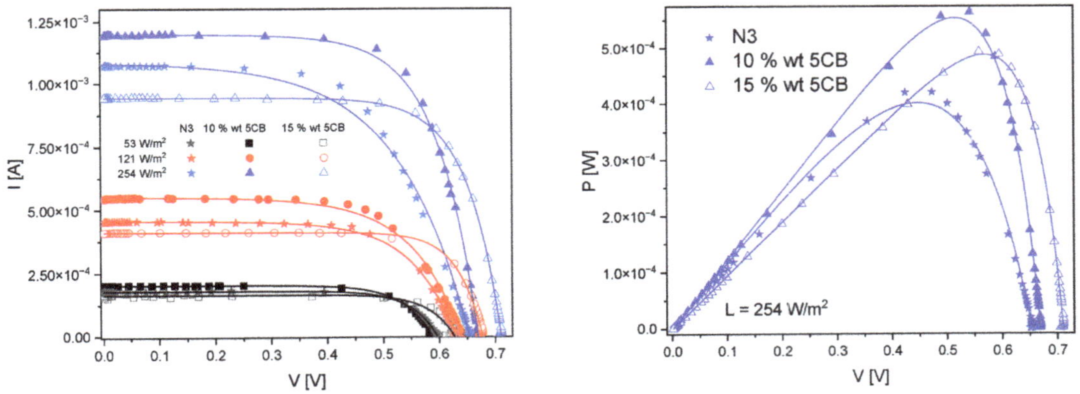

(a) (b)

Figure 7. (**a**) I-V characteristics of the DSSC cell with N3 dye and electrolyte modified with 5CB; (**b**) dependence of the power of the DSSC cell on the voltage.

Table 6. Parameters of the DSSC cells for 5CB-electrolyte (L = 254 W/m^2).

Sample	I_{SC} [A]	V_{OC} [V]	P_0 [W]	P_{max} [W]	FF	η [%]	I_{ph} [A]	I_0 [A]	n
N3	1.08×10^{-3}	0.652	7.04×10^{-4}	4.2×10^{-4}	0.60	4.6	$(1.080 \pm 0.004) \times 10^{-3}$	$(5.45 \pm 0.61) \times 10^{-5}$	8.1 ± 0.2
10% wt 5CB	1.20×10^{-3}	0.667	8.01×10^{-4}	5.61×10^{-4}	0.70	6.1	$(1.200 \pm 0.004) \times 10^{-3}$	$(8.76 \pm 1.26) \times 10^{-8}$	14.3 ± 0.2
15% wt 5CB	0.94×10^{-3}	0.709	6.69×10^{-4}	4.89×10^{-4}	0.73	5.3	$(0.944 \pm 0.004) \times 10^{-3}$	$(6.33 \pm 1.80) \times 10^{-9}$	16.8 ± 0.4

For a 10% wt 5CB concentration in the electrolyte, the I_{SC} is greater than both the I_{SC} in the unmodified cell and for the cell filled with an electrolyte with a 5CB concentration of about 15% wt (Table 6). Instead, for a cell with a 15% wt concentration of 5CB in the electrolyte, the I-V characteristic shows I_{SC} lower than both previously discussed characteristics. The obtained results suggest that at not too high concentrations of 5CB in the electrolyte, the presence of elongated molecules promotes increased transport of charge carriers as a result of the formation of conduction channels facilitating ion migration [17,47]. It is also possible that 5CB molecules interact with the TiO_2 surface, forcing their homeotropic orientation, which may facilitate carrier mobility in this region. The result of this process is an increase in the I_{SC}. On the other hand, increasing the concentration of polar molecules in the electrolyte can inhibit ion transport. The interaction of polar molecules and ions can result in a decrease in their mobility [48,49]. Moreover, excessive LC concentrations near the photoanode can block carrier flow and reduce the DSSC efficiency. This translates into a reduction in I_{SC}, even compared to cells with an unmodified electrolyte. It has been observed that as the alkyl chain length increases from 5CB to 10CB, ionic conductivity decreases. This is because the longer the alkyl chain, the better the insulating properties of the material. Long alkyl chains can impede ionic conductivity due to their entanglement. Furthermore, shorter alkyl chains result in lower viscosity of the LC material [49].

However, as 5CB concentration increases, the increase in V_{OC} is observed—the greater the LC concentration, the greater the value of V_{OC} (Table 6). This may be due to enhanced molecular packing in the active layer and increased charge dissociation. Interestingly, the FF also increases with the increase in concentration. This is probably the effect of changes in both I_{SC} and V_{OC}, whereas the η of converting light energy into electric current changes depending on LC concentration and is the largest for some optimal concentration, in this case 10% wt 5CB. A greater spreading of the I-V characteristics is also observed with the increase in light intensity. This is the result of the increased conversion of photons to electrons as a result of the increased efficiency of this process in the dye. The described results require further detailed studies using LCs with different alkyl chain lengths (e.g., homologous series of n-cyanobiphenyls) or forming other LC phases, e.g., smectic or columnar phases. It should be noted that increasing the light intensity can lead to a change in the temperature at which the measurement is performed. Measurements should be performed under stable temperature conditions and within a short time frame to minimize the impact on measurement results. The influence of temperature on photovoltaic cell parameters is a well-known issue [50,51].

4. Conclusions

The I-V characteristics obtained for DSSC dye cells show typical features for photovoltaic cells. The observed I_{SC} are small, of the order of several or tenths of a milliampere, while the V_{OC} reach values of the order of 0.7 V. The parameters I_{SC}, U_{OC}, P_0, and P_{max} change proportionally to the changes in the light intensity. However, the parameter FF slightly increases with decreasing light intensity, but η shows some kind of maximum. The obtained value of the ideality factor n for all samples differs significantly from the theoretical value predicted by the Shockley–Read–Hall recombination theory, which suggests that current does not flow uniformly throughout the cells, but always in local sites, e.g., in extended defects. As the dye concentration increases, the I_{SC} also increases, and it is crucial that the changes in dye concentration are neither too large nor too small. This significantly affects the efficiency of DSSC cells. The type of dye and its molecular structure are important for the efficiency of the tested DSSC cells. The measurements showed that the highest efficiency was achieved by DSSC cells sensitized with the N3 dye. In contrast, the remaining two dyes, N719 and Z907, were characterized by a lower level of light conversion.

Adding the nematic LC 5CB to the iodide electrolyte significantly modifies the efficiency of the DSSC cell. It was observed that for the lower concentrations of 5CB in the electrolyte (of the order of 10% wt), the *Isc*, and consequently the efficiency of the cell, increases. However, higher concentrations of 5CB in the electrolyte lead to a decrease in efficiency due to the inhibition of charge carriers' mobility. The highest value of 6.1% was obtained for DSSC cells containing 10% wt 5CB in the electrolyte. Also, the presence of terminal cyano-groups of 5CB increases the intermolecular interactions with acetonitrile molecules, which improve the thermal stability of the electrolyte. Further studies are needed to find the optimal LC concentrations, the influence of the type and structure of mesophases (smectic phase, discotic phase, etc.), and the molecular structure of LC molecules (e.g., alkyl chain length) on the conversion efficiency in DSSC cells. The obtained results will allow for the optimization of the preparation and operating conditions of DSSC cells.

Author Contributions: Conceptualization, S.A.R.; Methodology, S.A.R.; Software, P.S.; Formal Analysis, S.A.R.; Investigation, P.S.; Data Curation, P.S.; Writing—Original Draft, S.A.R.; Project Administration, P.S. All authors have read and agreed to the published version of the manuscript.

Funding: This research received no external funding.

Data Availability Statement: Data are contained within the article.

Acknowledgments: The authors would like to thank the GWDA sp. z o.o., Piła, Poland, for providing the renewable energy sources laboratory.

Conflicts of Interest: The authors declare no conflicts of interest.

References

1. Ang, T.-Z.; Salem, M.; Kamarol, M.; Das, H.S.; Nazari, M.A.; Prabaharan, N. A comprehensive study of renewable energy sources: Classifications, challenges and suggestions. *Energy Strateg. Rev.* **2022**, *43*, 100939. [CrossRef]
2. Al-Ezzi, A.S.; Ansari, M.N.M. Photovoltaic solar cells: A review. *Appl. Syst. Innov.* **2022**, *5*, 67. [CrossRef]
3. Schropp, R.E.I.; Carius, R.; Beaucarne, G. Amorphous silicon, microcrystalline silicon, and thin-film polycrystalline silicon solar cells. *MRS Bull.* **2007**, *32*, 219–223. [CrossRef]
4. Nakajima, K.; Usami, N. (Eds.) *Crystal Growth of Si for Solar Cells*; Springer: Berlin/Heidelberg, Germany, 2009. [CrossRef]
5. Battaglia, C.; Cuevas, A.; De Wolf, S. High-efficiency crystalline silicon solar cells: Status and perspectives. *Energy Environ. Sci.* **2016**, *9*, 1552–1576. [CrossRef]
6. Konagai, M. Present status and future prospects of silicon thin-film solar cells. *Jpn. J. Appl. Phys.* **2011**, *50*, 030001. [CrossRef]
7. Pandey, A.K.; Shahabuddin, S.; Ahmad, M.S. (Eds.) *Dye-Sensitized Solar Cells*; Elsevier Academic Press: London, UK, 2022.
8. Rawal, N.; Vaishaly, A.G.; Sharma, H.; Mathew, B.B. Dye sensitized solar cells: The emerging technology. *Energy Power Eng. Sci.* **2015**, *2*, 46–52. [CrossRef]
9. Hoppe, H.; Sariciftci, N.S. Organic solar cells: An overview. *J. Mater. Res.* **2004**, *19*, 1924–1945. [CrossRef]
10. Andualem, A.; Demiss, S. Review on dye-sensitized solar cells (DSSCs). *J. Heterocycl.* **2018**, *1*, 29–34. [CrossRef]
11. Bera, S.; Sengupt, D.; Roy, S.; Mukherjee, K. Research into dye-sensitized solar cells: A review highlighting progress in India. *J. Phys. Energy* **2021**, *3*, 032013. [CrossRef]
12. Uchida, J.; Soberats, B.; Gupta, M.; Kato, T. Advanced functional liquid crystals. *Adv. Mater.* **2022**, *34*, 2109063. [CrossRef]
13. Kato, T.; Uchida, J.; Ichikawa, T.; Sakamoto, T. Functional liquid crystals towards next generation of materials. *Angew. Chem.* **2018**, *57*, 4355–4371. [CrossRef]
14. Lagerwall, J.P.F.; Scalia, G. A new era for liquid crystal research: Applications of liquid crystals in soft matter. *Curr. Appl. Phys.* **2012**, *12*, 1387–1412. [CrossRef]
15. Bisoyi, H.K.; Kumar, S. Liquid-crystal nanoscience: An emerging avenue of soft self-assembly. *Chem. Soc. Rev.* **2011**, *40*, 306–319. [CrossRef]
16. Smaisim, G.F.; Mohammed, K.J.; Hadrawi, S.K.; Koten, H.; Kianfar, E. Properties and application of nanostructure in liquid crystals: Review. *BioNanoScience* **2013**, *13*, 819. [CrossRef]
17. Huang, C.Y.; You, C.F.; Cheng, C.E.; Lei, B.C.; Jhang, J.C.; Yu, F.C.; Chang, C.S.; Sen Chien, F.S. Liquid crystal-doped liquid electrolytes for dye sensitized solar cell applications. *Opt. Mater. Express* **2016**, *6*, 1024–1031. [CrossRef]
18. Ahmad, Z.; Hong, Z.; Viswanathan, V. Design rules for liquid crystalline electrolytes for enabling dendrite-free lithium metal batteries. *Proc. Natl. Acad. Sci. USA* **2020**, *117*, 26672–26680. [CrossRef]

19. Sasi, R.; Sarojam, S.; Devaki, S.J. High performing biobased ionic liquid crystal electrolytes for supercapacitors. *ACS Sustain. Chem. Eng.* **2016**, *4*, 3535–3543. [CrossRef]
20. Ruan, Q.; Yao, M.; Yuan, D.; Dong, H.; Liu, J.; Yuan, X.; Fang, W.; Zhao, G.; Zhang, H. Ionic liquid crystal electrolytes: Fundamental, applications and prospects. *Nano Energy* **2023**, *106*, 108087. [CrossRef]
21. Devaki, S.J.; Sasi, R. Ionic liquids/ionic liquid crystals for safe and sustainable energy storage systems. In *Progress and Developments in Ionic Liquids*; Handy, S., Ed.; InTechOpen: London, UK, 2017; pp. 313–336. [CrossRef]
22. Moisés, I.A.; Innocenti, A.; Somville, M.; Notredame, B.; Passerini, S.; Gohy, J.-F. Liquid crystals as additives in solid polymer electrolytes for lithium metal batteries. *MRS Adv.* **2023**, *8*, 797–802. [CrossRef]
23. Ghosh, S.; Ramos, L.; Remita, H. Swollen hexagonal liquid crystals as smart nanoreactors: Implementation in material chemistry for energy applications. *Nanoscale* **2018**, *10*, 5793–5819. [CrossRef]
24. BinSabt, M.H.; Atta, N.F.; Ahmed, Y.M.; Galal, A. Nanostructured Pt-Ru /ionic liquid crystal composite for electrocatalytic oxidation of methanol. *Int. J. Electrochem. Sci.* **2017**, *12*, 11271–11286. [CrossRef]
25. Ana, Y.; Han, X.; Liu, Y.; Azhar, A.; Na, J.; Nanjundan, A.K.; Wang, S.; Yu, J.; Yamauchi, Y. Progress in solid polymer electrolytes for lithium-ion batteries and beyond. *Nano-Micro Small* **2022**, *18*, 2103617. [CrossRef]
26. Kapernaum, N.; Lange, A.; Ebert, M.; Grunwald, M.A.; Haege, C.; Marino, S.; Zens, A.; Taubert, A.; Giesselmann, F.; Laschat, S. Current topics in ionic liquid crystals. *ChemPlusChem* **2022**, *86*, e202100397. [CrossRef]
27. Lu, F.; Gao, X.; Dong, B.; Sun, P.; Sun, N.; Xie, S. Nanostructured proton conductors formed via in situ polymerization of ionic liquid crystals. *ACS Appl. Mater. Interfaces* **2014**, *6*, 21970–21977. [CrossRef]
28. Salikolimi, K.; Sudhakar, A.A.; Ishida, Y. Functional ionic liquid crystals. *Langmuir* **2020**, *36*, 11702–11731. [CrossRef] [PubMed]
29. Högberg, D.; Soberats, B.; Uchida, S.; Yoshio, M.; Kloo, L.; Segawa, H.; Kato, T. Nanostructured two-component liquid-crystalline electrolytes for high-temperature dye-sensitized solar cells. *Chem. Mater.* **2014**, *26*, 6496–6502. [CrossRef]
30. Högberg, D.; Soberats, B.; Yoshio, M.; Mizumura, Y.; Uchida, S.; Kloo, L.; Segawa, H.; Kato, T. Self-assembled liquid-crystalline ion conductors in dye-sensitized solar cells: Effects of molecular sensitizers on their performance. *ChemPlusChem* **2017**, *82*, 834–840. [CrossRef]
31. Chen, R.; Weng, Q.; An, Z.; Zhu, S.; Wang, Q.; Chen, X.; Chen, P. Investigation of 4-pyridyl liquid crystals on the photovoltaic performance and stability of dye sensitized solar cells by the co-sensitization. *Dye. Pigm.* **2018**, *159*, 527–532. [CrossRef]
32. Solaronix Materials. Available online: www.solaronix.com (accessed on 24 July 2025).
33. Hossain, M.K.; Pervez, M.F.; Mia, M.N.H.; Mortuz, A.A.; Rahaman, M.S.; Karim, M.R.; Islam, J.M.M.; Ahmed, F.; Khan, M.A. Effect of dye extracting solvents and sensitization time on photovoltaic performance of natural dye sensitized solar cells. *Results Phys.* **2017**, *7*, 1516–1523. [CrossRef]
34. Krawczak, E.; Zdyb, A. The effect of electrode immersion time and ageing on N719 dye-sensitized solar cells performance. *J. Ecol. Eng.* **2020**, *21*, 53–60. [CrossRef]
35. Bakr, N.A.; Ali, A.R.K.; Jassim, S.M.; Hasoon, K.I. Effect of N719 dye concentration on the conversion efficiency of Dye Sensitized Solar Cells (DSSCs). *Zanco J. Pure Appl. Sci.* **2017**, *29*, 274–280.
36. Rahman, M.Y.A.; Umar, A.A.; Roza, L.A.; Salleh, M.M. Effect of organic dye on the performance of dye sensitized solar cell utilizing TiO_2 nanostructure films synthesized via CTAB-assisted liquid phase deposition technique. *Russ. J. Electrochem.* **2014**, *50*, 1072. [CrossRef]
37. Thoms, E.; Yu, L.; Richert, R. From very low to high fields: The dielectric behavior of the liquid crystal 5CB. *J. Mol. Liq. A* **2022**, *368*, 120664. [CrossRef]
38. Chandrasekar, V.; Lu, R.J.; Dierking, I. Micro-scale viscosity measurements of different thermotropic and lyotropic classes of liquid crystals by using ferrofluid inclusions. *J. Mol. Liq.* **2023**, *383*, 122178. [CrossRef]
39. Tamrakar, V.; Gupta, S.C.; Yashwant, S. Single-diode PV cell modeling and study of characteristics of single and two-diode equivalent circuit. *Int. J. Electr. Electron. Eng.* **2015**, *4*, 13–24. [CrossRef]
40. Zhang, C.; Zhang, J.; Hao, Y.; Lin, Z.; Zhu, C. A simple and efficient solar cell parameter extraction method from a single current-voltage curve. *J. Appl. Phys.* **2011**, *110*, 064504. [CrossRef]
41. Ishibashi, K.; Kimura, Y.; Niwano, M. An extensively valid and stable method for derivation of all parameters of a solar cell from a single current-voltage characteristic. *J. Appl. Phys.* **2008**, *103*, 094507. [CrossRef]
42. de Andrade, R.L.; de Oliveira, M.C.; Kohlrausch, E.C.; Santos, M.J.L. Simplified and quick electrical modeling for dye sensitized solar cells: An experimental and theoretical investigation. *J. Phys. Chem. Solids* **2018**, *116*, 273–280. [CrossRef]
43. Chegaar, M.; Hamzaoui, A.; Namoda, A.; Petit, P.; Aillerie, M.; Herguth, A. Effect of illumination intensity on solar cells parameters. *Energy Procedia* **2013**, *36*, 722–729. [CrossRef]
44. Różański, S.A. Determination of the current-voltage characteristics of the photovoltaic cells using the CoachLabII+ measuring console. *J. Autom. Electron. Electr. Eng.* **2022**, *4*, 31–38. [CrossRef]
45. Breitenstein, O. Understanding the current-voltage characteristics of industrial crystalline silicon solar cells by considering inhomogeneous current distributions. *Opto-Electron. Rev.* **2013**, *21*, 259–282. [CrossRef]

46. Rahman, M.Y.A.; Umar, A.A.; Taslim, R.; Salleh, M.M. Effect of organic dye, the concentration and dipping time of the organic dye N719 on the photovoltaic performance of dye-sensitized ZnO solar cell prepared by ammonia-assisted hydrolysis technique. *Electrochim. Acta* **2013**, *88*, 639–643. [CrossRef]
47. Zhu, J.S.; Su, Y.; Liu, Y.; Niu, X.; Zhang, W.; Chen, W. Liquid crystals as electrolyte additives for improving the transmission paths of redox couples and performances of dye-sensitized solar cells. *Opt. Mater.* **2024**, *149*, 115007. [CrossRef]
48. You, C.-F.; Cheng, C.-E.; Lei, B.-C.; Jhang, J.-C.; Yu, F.-C.; Chang, C.-S.; Chien, F.S.-S.; Huang, C.-Y. Improvement of the efficiency of dye-sensitized solar cells with fluorinated carbon-based liquid crystal dopant. *Chin. J. Phys.* **2017**, *55*, 1189. [CrossRef]
49. Kamarudin, M.A.B. Integration of Liquid Crystals with Redox Electrolytes in Dye-Sensitised Solar Cells. Ph.D. Thesis, University of Cambridge, Cambridge, UK, 2017.
50. Lee, D.K.; Suresh, T.; Yun, H.J.; Kim, Y.R.; Kim, J.H. Effect of temperature on the photovoltaic performance of dye-sensitized solar cells based on an iodine quasi-solid state electrolyte. *Sci. Adv. Mater.* **2016**, *8*, 1116–1121. [CrossRef]
51. Raga, S.R.; Fabregat-Santiago, F. Temperature effects in dye-sensitized solar cells. *Phys. Chem. Chem. Phys.* **2013**, *15*, 2328–2336. [CrossRef]

crystals

MDPI

Article

Light-Induced Heating of Microsized Nematic Volumes

Dmitrii Shcherbinin [1,*], Denis A. Glukharev [1], Semyon Rudyi [1], Anastasiia Piven [2], Tetiana Orlova [3], Izabela Śliwa [4] and Alex Zakharov [5]

1 International Research and Educational Centre for Physics of Nanostructures, ITMO University, St. Petersburg 197101, Russia; daglukharev@itmo.ru (D.A.G.); ssrudyi@itmo.ru (S.R.)
2 Infochemistry Scientific Center, ITMO University, St. Petersburg 197101, Russia; aopiven@itmo.ru
3 Institute of Physics, Yerevan State University, Yerevan 0025, Armenia; tetiana.orlova@ysu.am
4 Department of Operations Research and Mathematical Economics, Poznan University of Economics and Business, 61-875 Poznan, Poland; izabela.sliwa@ue.poznan.pl
5 Institute for Problems in Mechanical Engineering, Russian Academy of Science (IPME RAS), St. Petersburg 199178, Russia; alexandre.zakharov@yahoo.ca
* Correspondence: dpshcherbinin@itmo.ru

Abstract

The experimental study has been carried out using advanced computer vision methods in order to visualize the moment of excitation and further propagation of a non stationary isotropic domain in a hybrid aligned nematic (HAN) microsized volume under the effect of a laser beam focused on a bounding liquid crystal surface. It has been shown that, when the laser power exceeds a certain threshold value, in bulk of the HAN microvolume, an isotropic circular domain is formed. We also observed a structure of alternating concentric rings around the isotropic circular region, which increases with distance from the center of the isotropic domain. The formation of a sequence of rings in a polarizing microscopic image indicates the formation of a complex topology of the director field in the HAN cell under study. The following evolution of the texture can be represented by two modes. Firstly, the "fast" heating mode, which is responsible for the formation and explosive expansion of an isotropic zone in bulk of the HAN microvolume with characteristic time τ_1 due to a laser spot heating on the upper indium tin oxide (ITO) layer. Secondly, the "slow" heating mode, when an isotropic zone and concentric rings slowly expand with characteristic time τ_2 mainly due to the finite thermoconductivity of ITO layer. When the laser power significantly exceeds the threshold value, damped oscillations of the isotropic domain are observed. We also introduced the metrics that allows quantitatively estimate the behavior of texture observed. The results obtained form an experimental basis for further investigation of thermomechanical force appearing in the LC system with coupled gradients of temperature and director fields.

Keywords: liquid crystals; hybrid alignment nematic; computer vision; thermomechanical force; phase transition

PACS: 61.30.Cz; 64.70.Md

1. Introduction

With the current sustained demand for further miniaturization in the drug delivery devices, as well as techniques for biomolecules manipulation and biosensing, there is considerable interest in understanding the mentioned systems behavior from a technological implementation point of view. Robust and accurate experimental techniques for

https://doi.org/10.3390/cryst15090822

controlling the movement of molecular liquids and liquid crystals (LCs) make it possible to quickly and cost-effectively simulate hypothetical devices. This, in turn, can lead to the development of new and potentially improved designs for biotechnology devices and approaches. Thus, the investigation of complex dynamics in organic fluids including anisotropic materials is still of a high demand of modern soft matter physics. Microfluidics has appeared as a scientific field which is devoted to the approach how to manipulate, control and analyses tiny amounts of molecular liquids and LCs in channels ranging in size from tens to hundreds of micrometers [1–3]. Microfluidics has become a paradigm in various fields from chemical synthesis and biological analysis to optics and information technology [1,4,5]. Central to the success in microfluidics is development of innovative methods of manipulation of molecular liquids or LC systems in microchannels. Complex fluids, such as LCs or systems with biological materials, are characterized by a non-trivial coupling between fluid flow and structural deformations [4–8]. The channel size up to nanoscale ensures a close connection between nematic and biological fluids [4]. As for LC, an external electric field is traditionally used for controlling fluid motion [1]. On the other hand, in the significantly reduced LC volume, a number of other factors affect the flow dynamics. Among those factors is the temperature gradient, which starts to play an important role.

A liquid crystal drop of micro- sized volume is extremely sensitive to both the temperature gradient, ∇T, and boundary conditions. Consequently, these factors can influence the nature of hydrodynamic flow $\mathbf{v}$ in micro- or nanosized LC channels. One of the nonmechanical methods for producing flow in a nematic microfluidic channel is based on the coupling between the temperature, ∇T , and the director field, $\nabla \hat{\mathbf{n}}$, gradients. The problem of motion associated with an ultra-small (a few microliters) isotropic and LC drop, under the influence of the temperature gradient, caused by a laser beam, has drawn considerable attention [9–13]. The possibility of using a nearly infrared laser as a microfluidic heat source has been addressed by several groups [9–13]. Lasers provide the opportunity of precise tunable spatial and temporal properties of optical irradiation, providing controllable heating of the LC system [13]. Understanding of how the liquid crystal material can be manipulating by the temperature gradient is also a matter of great fundamental interest, as well as an essential part of knowledge in the field of soft-matter science.

If the laser beam is focused on the volume of the LC droplet or channel, as well as on the bounding surfaces, then an isotropic domain can form in the volume of the LC phase, or near the bounding surfaces, for some long time. Coupling between ∇T and $\nabla \hat{\mathbf{n}}$ is responsible for the excitation of an additional force in the LC phase called the thermomechanical force [14], (TMF) $\mathcal{F}_i = \partial \sigma_{\mathrm{ki}}^{\mathrm{TM}} / \partial x_k$. Here $\sigma_{\mathrm{ki}}^{\mathrm{TM}} = \xi_{\mathrm{kijlp}} (\nabla_j T)(\nabla_l n_p)$ is the thermomechanical stress tensor, which is one of the main factors that distinguish the LC system from isotropic molecular liquids, while ξ_{kijlp} is the thermomechanical coupling tensor and equal indexes are summed up. The magnitude of the hydrodynamic flow $\mathbf{v}$ excited by ∇T in a hybrid aligned nematic volume (HAN) is proportional to the tangential component of the thermomechanical stress tensor, $\sigma_{\mathrm{zx}}^{\mathrm{TM}}$, and has the form $\sim \frac{d}{\eta} \sigma_{\mathrm{zx}}^{\mathrm{TM}}$, where d and η are the thickness and viscosity of the liquid crystal material [8,9], respectively. Thus, the development of physically controlled systems is required to accurately study the thermomechanical force. As is known, the director gradient, $\nabla \hat{\mathbf{n}}$, can be formed by hybrid alignment of the LC volume on the bounding surfaces or near the isotropic domain-LC interface, or by taking into account both these mechanisms. At the same time, the temperature gradient ∇T can be excited by local laser heating. A recent analysis of the numerical results showed that due to the interaction of the gradients of the director's field, $\nabla \hat{\mathbf{n}}$ with ∇T, the heating pattern of the microscopic HAN cell in the direction perpendicular to the laser spot on the bounding surface (in the direction of z) is such that only a small part of the nematic volume is involved in the heating process, while most of the liquid crystal is not heated (see reference [15]). At the same time, this numerical investigation showed that the heating of the micro-sized volume of HAN caused by laser radiation spreads much wider in the xy-plane. Thus, heating of a micro-sized HAN cell in the xy-plane requires a detailed study of both the structure and dynamics of this system. For further development of concepts and models describing the thermomechanical force and energy dissipation modes in micro-volumes of HAN, we must find out, firstly, what typical dynamic modes can be observed in the HAN cell when exposed to a focused laser beam and, secondly, the typical evolution time of the director field disturbance. For this purpose, as the first step, an experimental study will be carried out using advanced computer vision methods that make it possible to capture the moment of excitation of the non-stationary isotropic domain in the HAN volume under the effect of temperature gradient. Thus, the main goal of our research is to develop experimental routines for a comprehensive study of the dynamics of formation of the isotropic domain that occurs in the nematic volume under laser radiation.

The outline of this paper is as follows: the experimental details are given in Section 2. Results and discussions are given in Section 3. Conclusions are summarized in Section 4.

2. Experimental Details

2.1. Sample Preparation

Let us begin with the consideration of the nematic sample exposed to laser radiation. The principal scheme of the LC cell is shown in Figure 1. To prepare LC cell, a standard indium tin oxide (ITO)-coated glass substrates are used. In our case, the ITO electrodes play the role of a light-absorbing layer, which is heated in the area of laser exposure. In the present study, we used the LC cell with a hybrid alignment. To provide a planar orientation on one of the bounding substrates, we used a polyimide alignment layer. A polyimide layer was deposited from solutions in 1-Methyl-2-pyrrolidinone (NMP, MACKLIN, Shanghai, China, ≥99%) to a pre-cleaned substrate, followed by spin-coating and drying at 220 °C for one hour. After drying, the polyimide layer was mechanically rubbed. To provide homeotropic anchoring on the ITO-coated glass plate, we used an adsorbed molecular monolayer of the surfactant CTAB (cetriltrimethylammonium bromide, Sigma-Aldrich, Saint Louis, MO, USA, ≥99%). The prepared substrates are fixed with glue. The thickness of a gap between substrates is set by teflon spacers. In the prepared cell, the thickness is equal to 15 μm , which is determined by the interference pattern in the transmission spectra of the cell before filling it with LC material. After the glue was completely dry, the LC cell was filled under the action of capillary forces. For the study, we use 4-pentyl-4′-cyanobiphenyl (5CB) nematic liquid crystals (Sigma Aldrich, 98%) with nematic-isotropic phase transition near 35 °C. In Figure 1 we schematically showed the heating of the upper ITO electrode by 1064 nm laser beam with following isotropic domain formation ("isotropic bubble", shown in grey color) and the initiation of a complex director field topology. In further discussion, we admit, that the plane of substrate is a xy-plane, the direction perpendicular to the substrate is along z-axis. The coordinate system is also shown in Figure 1.

2.2. Optical Setup and Measurement Routine

A developed experimental setup is shown in Figure 2. As a source of heating radiation we used a laser (VENUS, Connet Laser Technology, Shanghai, China) with a wavelength $\lambda_L = 1064$ nm in continuous mode (Laser 1064 in the Figure 2). To ensure the Gaussian distribution of the laser beam intensity profile, the spatial filter consisted of two lenses (L1, L2), as well as a pin-hole diaphragm (Ph) were used. Then, the laser beam after reflecting from the dichroic mirror (DM) is introduced into the objective (O), which focuses the beam on the upper electrode of the LC cell The objective used in the scheme allows ×50 magnification with a relatively longer working distance, which is 6 mm. The laser spot in the plane of focus (xy-plane) is approximately of 20 μm.

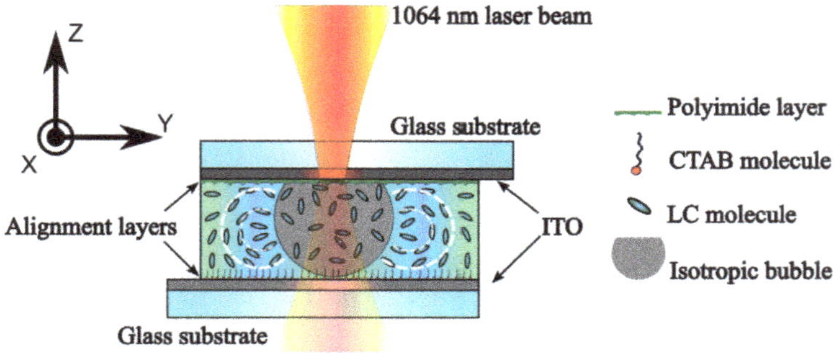

Figure 1. Schematic view of the HAN cell under the effect of focused laser beam.

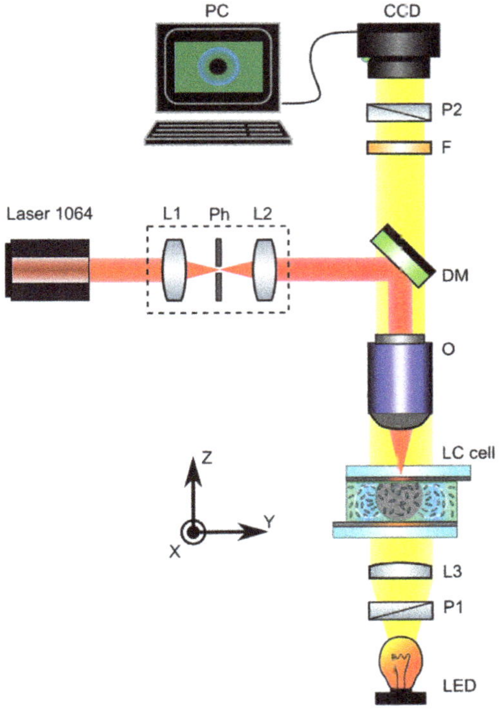

Figure 2. Top view on the experimental setup: P1 and P2 are the polarizers, L1, L2, and L3 are the lenses, F is the filter, O is the objective, DM is the dichroic mirror, and LED is the white light-emitting diode.

To obtain the texture of the LC layer, the polarized microscopic branch is used in the same setup. The white light from the LED source passes through the polarizer (P1, Figure 2). The linearly polarized light is focused on the LC cell using a lens (L3). The texture was observed using the same lens (O) and a CCD camera located after the color filter (F) and the cross analyzer (P2). The dichroic mirrors used have a high transmission coefficient in the visible spectrum and reflect in the near-infrared region of the spectrum. A CCD camera with a frame rate of 80 fps and a resolution of 640 × 480 pixels was used. We used a PC to collect the raw data (Figure 2).

Initially, we started by considering a HAN cell filled with 5CB molecules at a temperature of ∼300 K followed by exposure to a laser beam. The LC material was confined between two glass substrates with ITO layers and alignment layers. One variant of the hybrid aligned LC cell assumed that the upper bounding surface on which the laser beam is focused is aligned planar in the xy plane, while the lower bounding surface is aligned homeotropically, i.e., along z axis. This will be **case I** . Another variant of the hybrid aligned LC cell assumed that the upper bounding surface is aligned homeotropically, while the lower one is aligned planar. This will be **case II**. The LC cell with uniform thickness of 15 μm was placed horizontally (in xy-plane) in the optical setup (Figure 2). The tightly focused laser beam was used to effectively interact with the hybrid aligned 5CB sample through the ITO absorbing layer (see Figure 1). A number of heating regimes are considered, in which the laser radiation power P ranged from 0.5 to 15.5 mW with the corresponding laser energy exposition time $\tau_{\text{in}} = 8$ s. Here, we have to highlight one important aspect. In our setup using the HAN cell, focused laser radiation acts most effectively on the upper electrode. However, part of the de-focused transmitted laser light also interacts with lower electrode and can lead to "parasitic" heating of lower part of the LC cell. However, the temperature of the bottom electrode would be lower, which would allow the formation of the temperature gradient using laser radiation.

Thus, the measurement routine looks as follows:

1. The LC cell is arranged in the microscopic scheme between crossed polarizers. The position of the LC cell in the polarized optical microscopy branch is adjusted at 45 degrees to the plane of incident light polarization to ensure maximum transmittance before exposure to laser radiation.
2. First of all, we focus a low-power laser beam (1 MW) at the upper electrode. Secondly, we scan the cell position along the z axes. As will be shown in the next section, at low laser power, a stationary texture is quickly formed. During the z-scanning process, as the laser beam approaches the heated electrode, the texture begins to form. When focusing on the electrode, the texture will have the smallest size with the most pronounced intensity changes. After setting up the optical setup, the laser turns off to allow the LC cell to relax before measuring again.
3. Then we focus the laser on the electrode of the LC cell, which provides the planar alignment. In our work, we used several laser power levels from 0.5 mW to 15.5 mW.
4. For each certain power, we record a video footage of texture formation and evolution within 8 s after turning on the laser and 3 s before turning on the laser. For each power used, a full cycle of heating under the influence of the laser and cooling after the laser is turned off is performed.
5. The LC cell was rotated in the opposite direction. The laser radiation was focused on the electrode with the homeotropic boundary alignment. The video recording of texture formation was performed in accordance with the previous step.

3. Results and Discussion

3.1. Laser Beam Heating of Nematic Sample

Under the effect of the electrode heating with the laser power P exceeding a threshold value of approximately ≈2.5 mW, the "fast" formation of a texture consisting of a black circular region corresponding to an isotropic domain with concentric colored rings is observed. The typical texture is shown in Figure 3a for optical power equal to 15.5 mW. In Figure 3a we marked the center of texture with red dot and the revealed boundary of black area with dashed white line. To highlight the texture profile we provided an intensity cut section in Figure 4a at which there is clearly observed the central black area with following sequence of zones with high and low intensity.

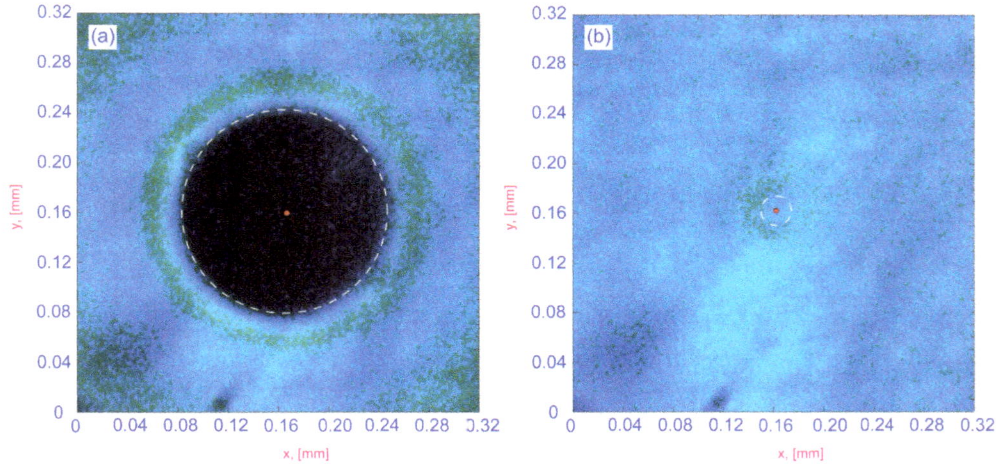

Figure 3. Formation of the isotropic domain in the HAN environment (**case I**) under the effect of laser radiation with the power of P = 15.5 mW (**a**) and the texture corresponding to the laser power below the threshold value, with P = 1.0 mW (**b**). Dashed circle represents the revealed features of the textures. The red dot corresponds to the center of the texture formed.

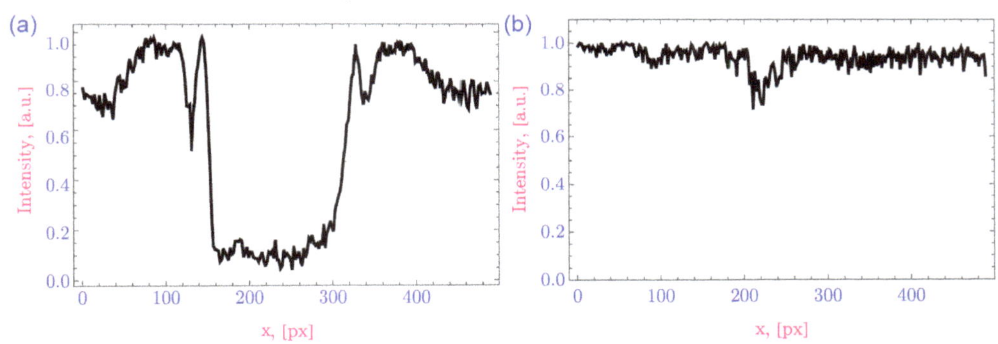

Figure 4. The intensity cut sections are shown in (**a**,**b**) for laser power equal to P = 15.5 mW and 1.0 mW, respectively.

We must emphasize that the textures are formed at a high speed, which apparently exceeds the frame rate used. However, after formation textures two modes of expansion are formed, namely the fast explosive expansion mode and the slow transition to a stationary state. The features of the evolution of the texture formed at the laser power exceeding the threshold volume will be discussed in the following subsections. Using the laser with the power below the threshold value (≈2.5 mW), we observed the formation of a slightly distinguishable colored ring without signs of an isotropic domain. The example of such a texture is shown in Figure 3b, where the laser power is P = 1.0 mW. The red dot corresponds to the center of the detected texture. A slightly observed colored ring is bounded by white lines. The corresponding intensity cut section is shown in Figure 4b.

Such a texture is completely formed at a rate much higher than the frame rate used per second, and can be characterized as static, since there is no sign of a change in the second time scale. As discussed in the previous section, we used this texture to provide focus on the electrode. The formation of the black domain in the xy-plane with a sequence of rings in the polarizing microscopic image indicates the formation of a complex topology of the director field in the HAN cell. This topology was caused by local heating and a corresponding change in the temperature gradient. Although these texture features are visible to the naked eye, and the radii of the observed rings can be estimated by the intensity of cut sections, we used computer vision methods to recognize the texture and analyze its morphology. The approach of computer vision techniques allow us to fast automatic process the obtained video-footage frame-by-frame and makes it possible to reduce the number of operator errors during manual processing. For this purpose, frame-by-frame preprocessing of the image was performed. At the first stage, we divided the source frame into RGB color channels and obtained 3 matrices with dimension 640×480 images containing information about the red, green and blue channels for each frame, respectively. For further data processing, it is sufficient to use only the blue channel matrices. In the second step, we removed the background matrix from each blue channel matrix, which corresponds to the averaging of blue channel data for frames taken 3 s before laser heating. At the next stage, each matrix was binarized and inverted for ease of further processing. Then, for each received preprocessed frame, a morphological closure operation was applied. The morphological detection algorithm was used to identify circular elements in each frame. The algorithm of morphological detection is described in Ref. [16]. The example of the preprocessing results is shown in Figure 5.

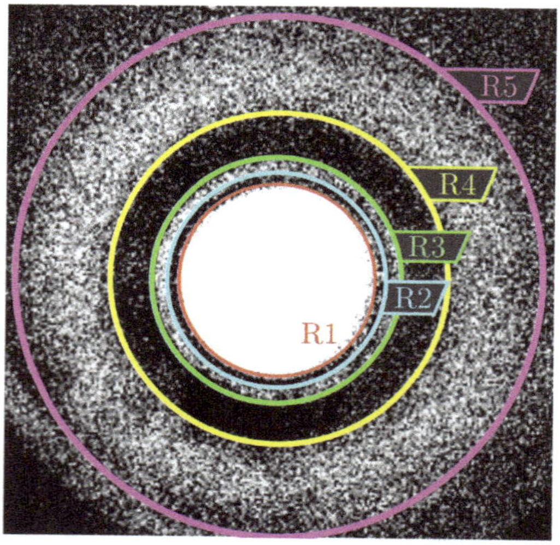

Figure 5. Analysis of HAN (**case I**) structures formed under the effect of laser radiation with the power of 15.5 mW. Radii RI represent the revealed boundaries of central circle (R1) and following concentric rings in the texture (R2–R5).

In the inverted binarized Figure 5, several circular elements can be observed when the circle with a radius R1 corresponds to the black isotropic area in Figure 3a. Other circular elements indicate the sequence of rings that appeared as a result of heating processes. Before we go any further, we need to clarify the physical nature of the texture formed in the polarized microscopic branch. Since the light from the LED propagates from bottom, perpendicular to the substrate, we represent the structure of the LC cell from the optical point of view as a series of vertically oriented zones, as shown in Figure 6.

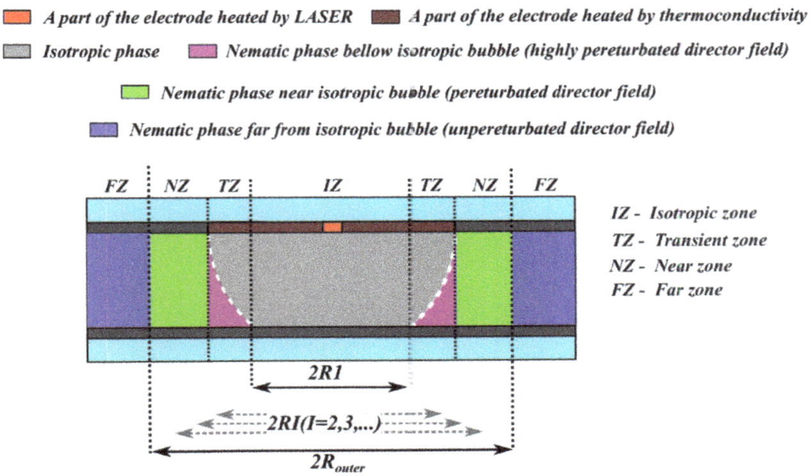

Figure 6. Formation of a sequence of an isotropic zone (IZ), a transient zone (TZ), a near zone (NZ), which have a perturbed director field but do not have an isotropic volume, and a far zone (FZ), which does not have a perturbed director field in the HAN cell under the influence of laser radiation.

To observe the black area in the polarized microscopic image (Figure 3a), the whole path from the bottom to the top of the substrate must correspond to a completely isotropic area. Thus, the black circular area in Figure 3a corresponds to the cylindrical volume of the isotropic domain, which contains only an isotropic liquid between substrates. Such a cylindrical zone is labeled as IZ in Figure 6. However, the total isotropic domain has a shape different from that of the cylinder. Thus, a part of the LC cell can be represented by a vertical zone containing both isotropic and nematic phases. Such a zone is designated as the transition zone (TZ) in Figure 6. Now it is clear that the radius of R1 from Figure 6 corresponds to the radius of the mentioned cylinder base, i.e., corresponds to the IZ-TZ boundary. We can assume that the director field in the nematic part of such zones can be strongly perturbed due to anchoring at the interface with an isotropic volume, possible fluid flows, and temperature gradient effects. Moreover, the thickness of the nematic phase in the TZ zone varies, as schematically shown in Figure 6. After TZ, we can introduce a zone called the near zone (NZ), with radius R_{outer}, which has a perturbed director field but no isotropic volume.

As seen in Figure 3a, in the texture the number of bright and dark rings are observed. From polarized optical microscopy point of view it means that the polarization of light after passing through the cell varied in the xy-plane. Indeed, after passing through a LC cell with induced complex director field topologies, phase transformations and hydrodynamic flows, the light from the LED is characterized by a complex spatial distribution of polarization depending on the local anisotropy. Thus, the textures carry information about the director field perturbation. Such a perturbation is attributed to changes in the director field topology, which determines the birefringence, and to changes in the nematic phase thickness, which determines the optical path through the anisotropic phases. The usage of crossed polarizers allows to transfer spatial topology of light polarization to the spatial topology of light intensity which is captured with CCD camera. Taking into account those thoughts, we can claim that all textural rings with radii RI starting from R2 are belonged to the TZ and NZ parts of cell between R_1 and R_{outer}, respectively (Figure 6).

Finally, the rest of the LC cell has no perturbations and can be designated as far zone (FZ). Since in FZ there is no spatial perturbation of director field and no isotropic phases, no textural features can be observed in FZ. Thus, the boundary between NZ and FZ (R_{outer} in

Figure 6) is an outer boundary of all textural features that can be observed. We can assume that number of rings as well as their thicknesses depend on following factors including thickness of the cell, initial configuration of director field, optical and material properties of LC used, parameters of laser irradiation. Further investigations may shed light on the influence of all these factors on formed texture properties.

Here we need to clarify another physical point. Despite the fact that the area of the laser spot on the upper ITO electrodes does not exceed ∼20 µm, the heating area of the electrode is significantly larger due to the limited thermal conductivity of the electrode materials, which is also shown in Figure 6. We would like to mention, that we have shown the typical textures observed for a given experimental condition; the reproducibility of texture formation features verified on several cells via measuring several times during a week.

3.2. Dynamics of the LC Textures

Now we can go further and focus our attention to the dynamics of texture evolution. As we said previously, the texture formed under the action of laser with power higher than ≈2.5 mW is not static and expands during the radiation exposure. First, we observe the "fast" mode characterized by explosive texture expansion over a typical period of time τ_1. We believe that this mode is mainly influenced by the heating of LC materials by a laser spot on the ITO layer and a less pronounced effect of thermal conductivity. Secondly, the "slow" mode, when the isotropic domain slowly increases due to the thermal conductivity of the electrode. This slow mode is characterized by a characteristic time period of τ_2.

As can be seen, the boundary between IZ and TZ is defined by a circle with radius R1. The first light ring in Figure 3a corresponds to the area of space between circles with radius R1 and R2 in the inverted binarized Figure 6, the second light ring corresponds to the area of space between the circle with radius R3 and R4, etc. All the bright ring structures observed in Figure 3a and characterized with R2, R3, etc are attributed to the TZ and NZ where the optical anisotropic properties of LC film is varying as discussed previously. When the laser is turned on, the observed textures and their radii expand. An example of the time dependence of the characteristic radii of a texture is shown in Figure 7 for **case I** and for optical laser power 15.5 mW.

The data presented in Figure 7 highlighted an explosive increase in the radius of the isotropic domain and the accompanying concentric rings, followed by a slow transition to a stationary state. The typical explosive growth time is about 0.1–0.2 s. When the laser radiation power is much higher than the threshold value, damped oscillations of the isotropic domain are observed.

Since, depending on the power of the laser radiation, the textures formed have different characteristic radii, and, generally speaking, the rings can extend beyond the frame, it is most advisable to compare the dynamics of texture expansion along the boundary of an isotropic zone. Figure 8 shows the dynamics of the isotropic zone expansion (the dependence of its radius $R1(t)$ on time t when the laser beam is focused on a planar (**case I**) (Figure 8a) and homeotropic (**case II**) (Figure 8b) substrates, depending on the power P of the laser radiation. As we can see from Figure 8a,b, the typical behavior of changing of the radius of zone IZ is the same for both **case I** and **case II**, consisting in explosive expansion and further slow growth. Moreover, damped oscillations are clearly observed at higher optical power values.

However, this phenomenon should be considered in detail in future work, by capturing the dynamics with higher frame rate and camera resolution, as well as by developing a hydrodynamic theoretical description.

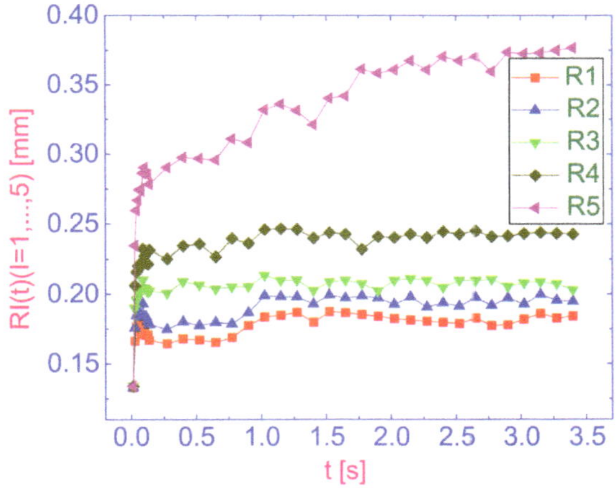

Figure 7. Evolution of the radius $RI(t)$ $(I = 1, \ldots, 5)$ of the circular textures in the HAN (**case I**) cell

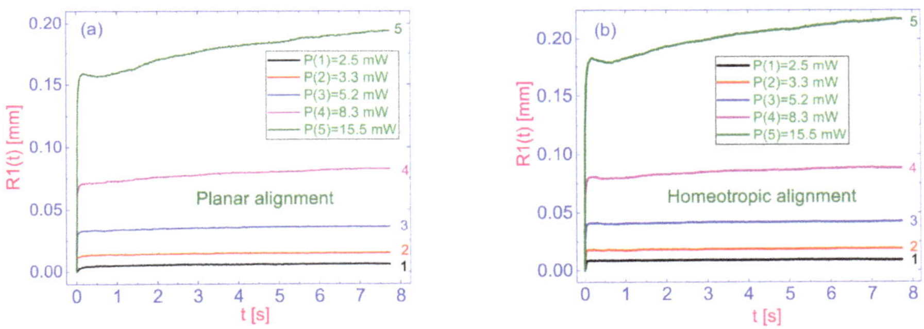

Figure 8. Evolution of the radius $R1(t)$ of the circular isotropic domain formed in the LC environment under the action of laser radiation of various power $P(i)$ $(i = 1, \ldots, 5)$, focused on the planar (**case I**) (**a**) and homeotropic (**case II**) (**b**) upper boundary, respectively.

At this stage, we intend to introduce some indicators that allow us to assess the dynamics of evolution more quantitatively. First, we need to estimate the stationary radii of the isotropic zones R_{max} that are formed when thermal stabilization in our dynamical system is achieved. To estimate R_{max}, we propose to plot the obtained dependencies $R1(t)$ in coordinates of $1/t$ for $t > 5$ s. The example of such plotting is shown in Figure 9a. The dependence of $R1(t)$ in this given time range t takes a linear form and can be extrapolated to the ordinate axis. The ordinate of the intersection point of such a linear approximation with the ordinate axis allows us to estimate the maximum radius of the circular domain R_{max}. The example of R_{max} determination is shown in Figure 9a.

Then, it is necessary to propose an algorithm for estimation of the characteristic time periods of the expansion dynamics. We can presume, that dynamics of expansion have multi-exponential behavior. Thus, the characteristic time can be obtained by plotting the time dependence of the IZ radii on logarithmic scale. For plotting convenience it is much easy to work with $\Delta R(t)$ which is $\Delta R1(t) = R_{max} - R1(t)$. As can be seen in Figure 9b linear dependence intervals can be observed using such a logarithmic scale when plotting $\ln(R_{max} - R1(t))$ on t. Two areas with linear dependencies are clearly distinguished: first, corresponding to the time just after switching on the laser radiation and, second, in the time interval between the 2 nd and 5 th seconds. The slope coefficients of the linear approximations correspond to the inverse of the characteristic times τ_1 and τ_2. Accordingly, the expected value of the first characteristic of the "fast" time, τ_1, is much less than the corresponding values of the "slow" time τ_2.

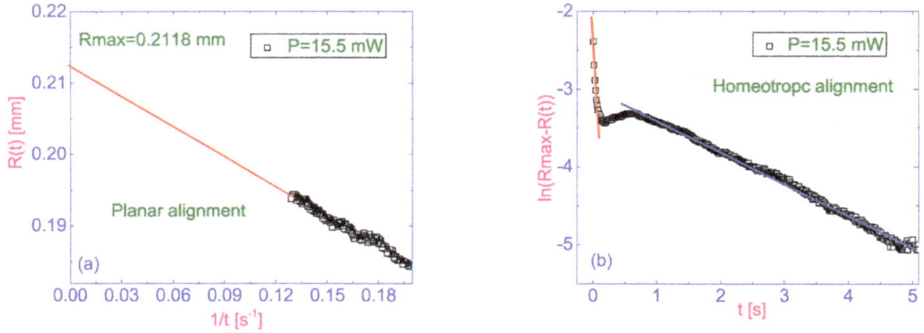

Figure 9. (**a**) An example of the determination of $R_{\max} \equiv R1$ by plotting in the inverse time coordinate with the following linear extrapolation. (**b**) The example of the determination of τ_1 and τ_2 by plotting in a logarithmic scale with the following linear approximations (red lines).

Comparison of the limiting radius $R_{\max}(P) \equiv R1$ of an isotropic zone depending on the power of P laser radiation when it focuses on both planar (**case I**) (triangles up) and homeotropically (**case II**) (down triangles) aligned substrates are shown in Figure 10. As can be seen from Figure 10, the two dependencies almost coincide. This dependence seems logical, since, as a result, the volume of the liquid that has passed into the isotropic phase should be determined by the amount of energy pumped into the system.

Figure 11 shows the dependences of "fast" ($\tau_1(P)$) and "slow" ($\tau_2(P)$) expansion time of the isotropic circular domain depending on the power P of the laser radiation when it is focused as on the planar, **case I**, (Figure 11a) and on homeotropic **case II**, (Figure 11b), substrates. In all cases, we see that τ_2 is several orders of magnitude higher than τ_1. Moreover, at this stage we have not identified a clear dependence of τ_2 of the laser power. In future work, we intend to obtain data with a smaller power step, which will allow us to investigate the dependence of τ_2 on the laser power and the initial configuration of the director's field. With a lower power P (less than 3 mW) and focusing on the planar aligned bounding surface (**case I**), a point appears that deviates sharply from the dependence $\tau_1(P)$. The observed flaw is caused by a technical error that occurred due to the limitation of the maximum time resolution of the CCD camera used (1/80 s). To improve the results of measuring the "fast" time τ_1, in the following studies it is necessary to reduce the power step and shoot at a higher frame rate, which will allow for a more accurate linear approximation. Nevertheless, starting with the power of P exceeding 3 mW, both dependencies tend to monotonous growth.

In Figure 12 we compare the dependence of the "fast" time τ_1 on the laser power for both **case I** and **case II**. It can be seen that the growth rates of these dependencies vary markedly. In the case of laser beam focusing on the homeotropically aligned bounding surface (**case II**), the dependence slope is almost twice as high as in the case of focusing on the planar aligned surface. This difference in the growth rates of these dependencies τ_1(**case I**) and τ_1(**case II**) can be explained by the difference in thermal conductivity coefficients $\lambda_{\parallel}$(5CB) ~ 0.24 and $\lambda_{\perp}$(5CB) ~ 0.13 (in units W/mK) [17] corresponding to the parallel and perpendicular direction of the director $\hat{\mathbf{n}}$.

Despite the fact that this stage of our experimental studies is the first attempt, we can assert that the presented techniques and metrics are sufficient for a comprehensively study of the dynamics of texture evolution in the context of the thermomechanical force in the LC cells with the gradient of both the director's field and temperature distribution.

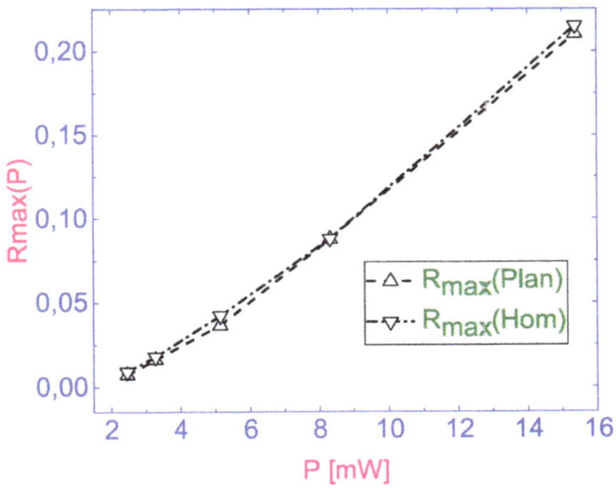

Figure 10. Dependencies the radius R (P) = R1(P) on the power P of the laser radiation.

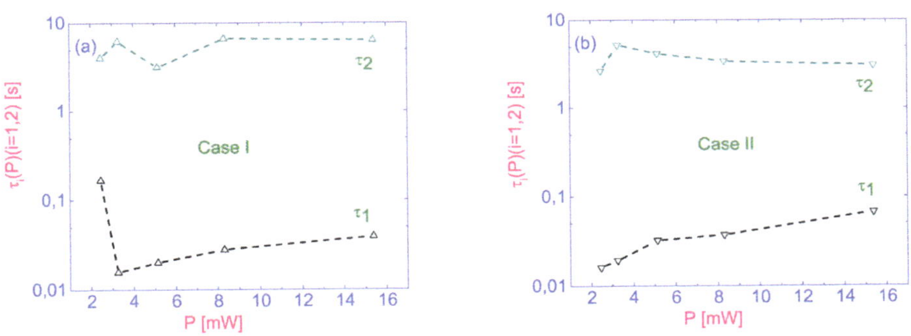

Figure 11. Dependencies of the "fast" $\tau_1(P)$ and "slow" $\tau_2(P)$ times on the power P of the laser radiation. In both these cases, the isotropic circular domain is surrounded by the HAN phase, with the nding surface.

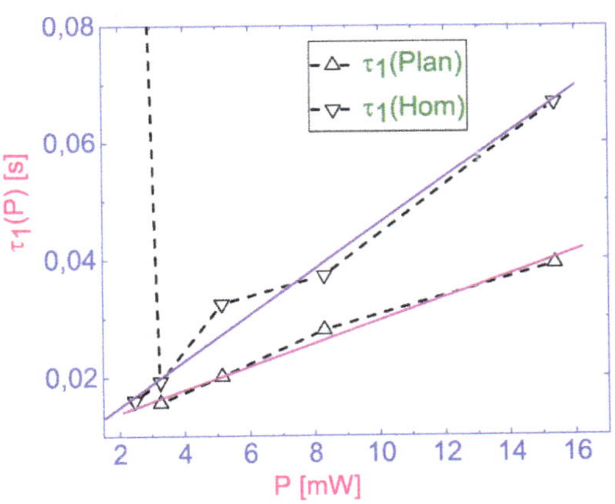

Figure 12. The comparison of $\tau_1(P)$ times dependencies on the power P of the laser radiation for **case I** and **case II**. Color lines indicate the trend.

4. Conclusions

Thus, we experimentally investigated the effect of laser heating on the formation and evolution of phase transformations in a hybrid aligned nematic cell (HAN). For this purpose, we have implemented an optical setup based on a polarized optical microscopic scheme with laser radiation injected. Then, an advanced computer vision method was used to visualize the moment of excitation and further propagation of an unsteady isotropic domain in the HAN micro-sized volume under the effect of a laser beam focused on the bounding LC (LC) surface. It has been shown that when the laser power exceeds a certain threshold value, in bulk of the HAN micro-volume an isotropic circular domain is formed. The structure of alternating concentric rings around the isotropic circular region was also observed, which increased with distance from the center of the isotropic domain. The formation of a sequence of rings in a polarizing microscopic image indicates the formation of a complex topology of the director field in the studied HAN cell. The following evolution of the texture can be represented by two modes. Firstly, the "fast" heating mode, which is responsible for the formation and explosive expansion of the isotropic zone in bulk of the HAN micro-volume with characteristic time τ_1 due to laser spot heating of the upper indium tin oxide (ITO) layer. It is shown that in the case of focusing the laser beam on the homeotropically aligned bounding surface (**case II**), the slope dependence of τ_1 is almost twice as high as in the case of focusing on the planar aligned surface (**case I**). Secondly, the "slow" heating mode, when the isotropic zone and concentric rings slowly expand with a characteristic time τ_2 mainly due to the finite thermoconductivity of the ITO layer. It is also shown that for both cases (**case I** and **case II**) the values τ_2 is an order of magnitude higher than τ_1. When the laser power significantly exceeds the threshold value, damped oscillations of the isotropic domain are observed. We have also introduced the metrics that allows us to quantify the behavior of the observed texture. The results obtained form the experimental basis for further investigation of the thermomechanical force appears in the LC system with coupled gradients of temperature and director fields. It should be noted that the study of the dynamics of laser heating in a microsized nematic volume consisting of cyanobiphenyl molecules can be extended to ferroelectric nematic materials [18,19].

We believe that the present study may shed some light on the problem of the micro-fluids dynamics in a micro-sized nematic volume under the heating effect of a focusing laser beam. The setup, image processing algorithms as well as metrics introduced in the present work may be become a basement of experimental routines for the investigation of thermomechanical force in LC cells stem from the coupling between the temperature and director field gradients. In turn, such studies will lead to the appearance of new microfluidic and micro cargo devices.

Author Contributions: D.S.: Writing—review and editing, Validation, Methodology, Investigation, Formal analysis, Conceptualization. S.R.: Writing—review and editing, Visualization, Software, Methodology, Investigation. D.A.G.: Investigation. A.P.: Investigation (as LC sample preparation). T.O.: Conceptualization, writing, review editing. I.Ś.: Conceptualization, writing, review editing. A.Z.: Conceptualization, Supervision, writing, review and editing. All authors have read and agreed to the published version of the manuscript.

Funding: A.Z. acknowledges financial support of the Ministry of Science and Higher Education of the Russian Federation for IPMash RAS (Grant Number FFNF-2024-0009).

Data Availability Statement: No data was used for the research described in the article; further inquiries can be directed to the corresponding author.

Conflicts of Interest: The authors declare no conflicts of interest.

References

1. Schoch, R.B.; Han, J.; Renaud, P. Transport phenomena in nanofluidics. *Rev. Mod. Phys.* **2008**, *80*, 839–883. [CrossRef]
2. Whitesides, G.M. The origins and the future of microfluidics. *Nature* **2006**, *442*, 368–373. [CrossRef] [PubMed]
3. Whitby, M.; Quirke, N. Fluid flow in carbon nanotubes and nanopipes. *Nat. Nanotechnol.* **2007**, *2*, 87–94. [CrossRef] [PubMed]

4. Rey, A.D. Liquid crystal models of biological materials and processes. *Soft Matter* **2010**, *6*, 3402–3429. [CrossRef]
5. Cuennet, J.B.; Vasdekis, A.E.; De Sio, L.; Psaltis, D. Optofluidic modulator based on peristaltic nematogen microflows. *Nat. Photonics* **2011**, *5*, 234–238. [CrossRef]
6. Woltman, S.J.; Jay, G.D.; Crawford, G.P. Liquid-crystal materials find a new order in biomedical applications. *Nat. Mater.* **2007**, *6*, 929–938. [CrossRef] [PubMed]
7. Squires T.M.; Quake, S.R. Microfluidics: Fluid physics at the nanoliter scale. *Rev. Mod. Phys.* **2005**, *77*, 977–1026. [CrossRef]
8. Śliwa, I.; Maslennikov, P.V.; Zakharov, A.V. Vortex Dynamics in a Hybrid Aligned Nematic Microvolume with an Orientational Defect. *Symmetry* **2023**, *15*, 324. [CrossRef]
9. Śliwa, I.; Zakharov, A.V. Heat driven flow in microsized nematic volumes: Computational studies and analysis. *Symmetry* **2021**, *13*, 459. [CrossRef]
10. Varneuil, E.; Cordero, M.L.; Gallaire, F.; Baroud, C. Laser-Induced Force on a Microfluidic Drop: Origin and Magnitude. *Langmuir* **2009**, *25*, 5127–5134. [CrossRef] [PubMed]
11. Zakharov, A.V.; Vakulenko, A.A. Thermally excited fluid flow in a microsized liquid crystal channel with a free surface. *Phys. Fluids* **2015**, *27*, 62001. [CrossRef]
12. Choi, H.; Takezoe, H. Circular flow formation triggered by Marangoni convection in nematic liquid crystal films with a free surface. *Soft Matter* **2016**, *12*, 481–485. [CrossRef] [PubMed]
13. Mao, H.; Arias-Gonszalez, J.R.; Smith, S.B.; Tinoco, I.; Bustamante, C. Temperature Control Methods in a Laser Tweezers System. *Biophys. J.* **2005**, *89*, 1308–1316. [CrossRef] [PubMed]
14. Akopyan, R.S.; Zeldovich, B.Y. Thermomechanical effects in deformed nematics. *Sov. Phys. JETP* **1984**, *87*, 1660–1669.
15. Śliwa, I.; Maslennikov, P.V.; Shcherbinin, D.P.; Zakharov A.V. Laser-driven nematic flow in microfluidic devices. *Phys. Rev.* **2024**, *E110*, 64702. [CrossRef] [PubMed]
16. Han, L.; Zhuang, Y.; Chen, K.; Xie, Y.; Liao, G.; Yin, G.; Lin, J. Circle Detection with Adaptive Parameterization: A Bottom-Up Approach. *Sensors* **2025**, *25*, 2552. [CrossRef] [PubMed]
17. Marinelli, M.; Ghosh, A.K.; Mercuri, F. Small quartz silica spheres induced disorder in octylcianobiphenyl (8CB) liquid crystals: A thermal study. *Phys. Rev.* **2001**, *E63*, 61713.
18. Chen, X.; Korblova, E.; Dong, D.; Wei, X.; Shao, R.; Radzihovsky, L.; Glaser, M.A.; Maclennan, J.E.; Bedrov, D.; Walb, D.M.; et al. First-principles experimental demonstration of ferroelectricity in a thermotropic nematic liquid crystal: Polar domains and striking electro-optics. *Proc. Natl. Acad. Sci. USA* **2020**, *117*, 14021–14031. [CrossRef] [PubMed]
19. Pan, J.-T.; Zhu, B.-H.; Ma, L.-L.; Chen, W.; Zhang, G.-Y.; Tang, J.; Liu, Y.; Wei, Y.; Zhang, C.; Zhu, Z.-H.; et al. Nonlinear geometric phase coded ferroelectric nematic fluids for nonlinear soft-matter photonics. *Nat. Commun.* **2024**, *15*, 8732. [CrossRef] [PubMed]

MDPI AG
Grosspeteranlage 5
4052 Basel
Switzerland
Tel.: +41 61 683 77 34

Crystals Editorial Office
E-mail: crystals@mdpi.com
www.mdpi.com/journal/crystals

www.ingramcontent.com/pod-product-compliance
Lightning Source LLC
Chambersburg PA
CBHW062053220926
51850CB00010B/2754
* 9 7 8 3 7 2 5 8 5 6 1 3 8 *